Classical Mechanics Simulations
The Consortium for Upper Level Physics Software

Bruce Hawkins
Department of Physics, Smith College,
Northampton, Massachusetts

Randall S. Jones
Department of Physics, Loyola College,
Baltimore, Maryland

Series Editors

Maria Dworzecka
Robert Ehrlich
William MacDonald

JOHN WILEY & SONS, INC.
NEW YORK · CHICHESTER · BRISBANE · TORONTO · SINGAPORE

ACQUISITIONS EDITOR Cliff Mills
MARKETING MANAGER Catherine Faduska
SENIOR PRODUCTION EDITOR Sandra Russell
MANUFACTURING MANAGER Susan Stetzer

This book was set in 10/12 Times Roman by Beacon Graphics and
printed and bound by Hamilton Printing Co. The cover was printed by Phoenix Color.

Recognizing the importance of preserving what has been written, it is a
policy of John Wiley & Sons, Inc. to have books of enduring value published
in the United States printed on acid-free paper, and we exert our best
efforts to that end.

The paper on this book was manufactured by a mill whose forest management programs include
sustained yield harvesting of its timberlands. Sustained yield harvesting principles ensure that
the number of trees cut each year does not exceed the amount of new growth.

Library of Congress Cataloging in Publication Data:
Hawkins, Bruce.
Classical mechanics simulations : the consortium for upper level
physics software / Bruce Hawkins, Randall S. Jones.
p. cm.
Includes bibliographical references (p.).
ISBN 0-471-54881-2 (pbk./disk)
1. Mechanics—Data processing. 2. Mechanics—Software
simulations. I. Jones, Randall S. II. Title.
QC125.2.H4 1995
530'.078—dc20 94-39444
CIP

Printed in the United States of America

10 9 8 7 6 5 4 3 2 1

Contents

List of Figures

1

Introduction

"It is nice to know that the computer understands the problem. But I would like to understand it too."

—Eugene P. Wigner, quoted in *Physics Today,* July 1993

1.1 Using the Book and Software

The simulations in this book aim to exploit the capabilities of personal computers and provide instructors and students with valuable new opportunities to teach and learn physics, and help develop that all-important, if somewhat elusive, physical intuition. This book and the accompanying diskettes are intended to be used as supplementary materials for a junior- or senior-level course. Although you may find that you can run the programs without reading the text, the book is helpful for understanding the underlying physics, and provides numerous suggestions on ways to use the programs. *If you want a quick guided tour through the programs, consult the "Walk Throughs" in Appendix A.* The individual chapters and computer programs cover mainstream topics found in most textbooks. However, because the book is intended to be a supplementary text, no attempt has been made to cover all the topics one might encounter in a primary text.

Because of the book's organization, students or instructors may wish to deal with different chapters as they come up in the course, rather than reading the chapters in the order presented. One price of making the chapters semi-independent of one another is that they may not be entirely consistent in notation or tightly cross-referenced. Use of the book may vary according to the taste of the student or instructor. Students may use this material as the basis of a self-study course. Some instructors may make homework assignments from the large number of exercises in each chapter or to use them as the basis of student projects. Other instructors may use the computer programs primarily for in-class demonstrations. In this latter case, you may find that the programs are suitable for a range of courses from the introductory to the graduate level.

Use of the book and software may also vary with the degree of computer programming performed by users. For those without programming experience, all the computer simulations have been supplied in executable form, permitting them to be used as is. On the other hand, Pascal source code for the programs has also been provided, and a number of exercises suggest specific ways the programs can be modified. Possible modifications range from altering a single procedure especially set up for this purpose by the author, to larger modifications following given examples, to extensive additions for ambitious projects. However, the intent of the authors is that the simulations will help the student to develop intuition and a deeper understanding of the physics, rather than to develop computational skills.

We use the term "simulations" to refer to the computer programs described in the book. This term is meant to imply that programs include complex, often realistic, calculations of models of various physical systems, and the output is usually presented in the form of graphical (often animated) displays. Many of the simulations can produce numerical output—sometimes in the form of output files that could be analyzed by other programs. The user generally may vary many parameters of the system, and interact with it in other ways, so as to study its behavior in real time. The use of the term simulation should not convey the idea that the programs are bypassing the necessary physics calculations and simply producing images that look more or less like the real thing.

The programs accompanying this book can be used in a way that complements, rather than displaces, the analytical work in the course. It is our belief that, in general, computational and analytical approaches to physics can be mutually reinforcing. It may require considerable analytical work, for example, to modify the programs, or really to understand the results of a simulation. In fact, one important use of the simulations is to suggest conjectures that may then be verified, modified, or proven false analytically. A complete list of programs is given in Section 1.7.

1.2 Required Hardware and Installation of Programs

The programs described in this book have been written in the Pascal language for MS-DOS platforms. The language is Borland/Turbo Pascal, and the minimum hardware configuration is an IBM-compatible 386-level machine preferably with math coprocessor, mouse, and VGA color monitor. In order to accommodate a wide range of machine speeds, most programs that use animation include the capability to slow down or speed up the program. To install the programs, place disk number 1 in a floppy drive. Change to that drive, and type Install. You need only type in the file name to execute the program. Alternatively, you could type the name of the driver program (the same name as the directory in which the programs reside), and select programs from a menu. A number of programs write to temporary files, so you should check to see if your autoexec.bat file has a line that sets a temporary directory, such as SET TEMP = C:\TEMP. (If you have installed WINDOWS on your PC, you will find that such a command has already been written into your autoexec.bat file.) If no such line is there, you should add one.

Compilation of Programs

If you need to compile the programs, it would be preferable to do so using the Borland 7.0 (or later) compiler. If you use an earlier Turbo compiler you may run out of memory when compiling. If that happens, try compiling after turning off memory resident programs. If your machine has one, be sure to compile with the math-coprocessor turned on (no emulation). Finally, if you recompile programs using any compiler other than Borland 7.0, you will get the message: "EGA/VGA Invalid Driver File" when you try to execute them, because the driver file supplied was produced using this version of the compiler. In this case, search for the file BGILINK.pas included as part of the compiler to find information on how to create the EGAVGA.obj driver file. *If any other instructions are needed for installation, compilation, or running of the programs, they will be given in a README file on the diskettes.*

1.3 User Interface

To start a program, simply type the name of the individual or driver program, and an opening screen will appear. All the programs in this book have a common user interface. Both keyboard and mouse interactions with the computer are possible. Here are some conventions common to all the programs.

Menus: If using the *keyboard*, press **F10** to highlight one of menu boxes, then use the **arrow** keys, **Home**, and **End** to move around. When you press **Return** a submenu will pull down from the currently highlighted menu option. Use the same keys to move around in the submenu, and press **Return** to choose the highlighted submenu entry. Press **Esc** if you want to leave the menu without making any choices.

If using the *mouse* to access the top menu, click on the menu bar to pull down a submenu, and then on the option you want to choose. Click anywhere outside the menus if you want to leave them without making any choice. Throughout this book, the process of choosing submenu entry **Sub** under main menu entry **Main** is referred to by the phrase "choose **Main | Sub**." The detailed structure of the menu will vary from program to program, but all will contain **File** as the first (left-most) entry, and under **File** you will find **About CUPS, About Program, Configuration,** and **Exit Program**. The first two items when activated by mouse or arrows keys will produce information screens. Selecting **Exit Program** will cause the program to terminate, and choosing **Configuration** will present you with a list of choices (described later), concerning the mode of running the program. In addition to these four items under the **File** menu, some programs may have additional items, such as **Open**, used to open a file for input, and **Save**, used to save an output file. If **Open** is present and is chosen, you will be presented with a scrollable list of files in the current directory from which to choose.

Hot Keys: Hot keys, usually listed on a bar at the bottom of the screen, can be activated by pressing the indicated key or by clicking on the hot key bar with the mouse. The hot key **F1** is reserved for help, the hot key **F10** activates the menu bar. Other hot keys may be available.

Sliders (scroll-bars): If using the *keyboard*, press **arrow** keys for slow scrolling of the slider, **PgUp/PgDn** for fast scrolling, and **End/Home** for moving from one end to another. If you have more then one slider on the screen then only the slider with marked "thumb" (sliding part) will respond to the above keys. You can toggle the mark between your sliders by pressing the **Tab** key.

If using the *mouse* to adjust a slider, click on the thumb of the slider, drag it to desired value, and release. Click on the arrow on either end of the slider for slow scrolling, or in the area on either side of thumb for fast scrolling in this direction. Also, you can click on the box where the value of the slider is displayed, and simply type in the desired number.

Input Screens: All input screens have a set of "default" values entered for all parameters, so that you can, if you wish, run the program by using these original values. Input screens may include circular radio buttons and square check boxes, both of which can take on Boolean, i.e., "on" or "off," values. Normally, check boxes are used when only one can be chosen, and radio buttons when any number can be chosen.

If using the *keyboard*, press **Return** to accept the screen, or **Esc** to cancel it and lose the changes you may have made. To make changes on the input screen by keyboard, use **arrow** keys, **PgUp**, **PgDn**, **End**, **Home**, **Tab**, and **Shift-Tab** to choose the field you want to change, and use the backspace or delete keys to delete numbers. For Boolean fields, i.e., those that may assume one of two values, use any key except those listed above to change its value to the opposite value.

If you use the *mouse*, click [OK] to accept the screen or [Cancel] to cancel the screen and lose the changes. Use the mouse to choose the field you want to change. Clicking on the Boolean field automatically changes its value to the opposite value.

Parser: Many programs allow the user to enter expressions of one or more variables that are evaluated by the program. The function parser can recognize the following functions: absolute value (abs), exponential (exp), integer or fractional part of a real number (int or frac), real or imaginary part of a complex number (re or im), square or square root of a number (sqr or sqrt), logarithms—base 10 or e (log or ln)—unit step function (h), and the sign of a real number (sgn). It can also recognize the following trigonometric functions: sin, cos, tan, cot, sec, csc, and the inverse functions arcsin, arccos, arctan, as well as the hyperbolic functions denoted by adding an "h" at the end of all the preceding functions. In addition, the parser can recognize the constants *pi*, *e*, $i(\sqrt{-1})$, and rand (a random number between 0 and 1). The operations **+**, **−**, *****, **/**, **^**(exponentiation), and **!**(factorial) can all be used, and the variables r and c are interpreted as $r = \sqrt{x^2 + y^2}$ and $c = x + iy$. Expressions involving these functions, variables, and constants can be nested to an arbitrary level using parentheses and brackets. For example, suppose you entered the following expression: **h(sin(10*pi*x))−0.5**. The parser would interpret this function as $h(sin(10\pi x)-0.5)$. If the program evaluates this function for a range of x-values, the result, in this case, would be a series of square pulses of width 1/15, and center-to-center separation 1/10.

Help: Most programs have context-sensitive help available by pressing the **F1** hot key (or clicking the mouse in the **F1** hot key bar). In some programs help is also available by choosing appropriate items on the menu, and in still other programs tutorials on various aspects of the program are available.

1.4 The CUPS Project and CUPS Utilities

The authors of this book have developed their programs and text as part of the Consortium for Upper-Level Physics Software (CUPS). Under the direction of the three editors of this book, CUPS is developing computer simulations and associated texts for nine junior- or senior-level courses, which comprise most of the undergraduate physics major curriculum during those two years. A list of the nine CUPS courses, and the authors associated with each course, follows this section. This international group of 27 physicists includes individuals with extensive backgrounds in research, teaching, and development of instructional software.

The fact that each chapter of the book has been written by a different author means that the chapters will reflect that individual's style and philosophy. Every attempt has been made by the editors to enhance the similarity of chapters, and to provide a similar user interface in each of the associated computer simulations. Consequently, you will find that the programs described in this and other CUPS books have a common look and feel. This degree of similarity was made possible by producing the software in a large group that shared a common philosophy and commitment to excellence.

Another crucial factor in developing a degree of similarity between all CUPS programs is the use of a common set of utilities. These CUPS utilities were written by Jaroslaw Tuszynski and William MacDonald, the former having responsibility for the graphics units, and the latter for the numerical procedures and functions. The numerical algorithms are of high quality and precision, as required for reliable results. CUPS utilities were originally based on the M.U.P.P.E.T. utilities of Jack Wilson and E.F. Redish, which provided a framework for a much expanded and enhanced mathematical and graphics library. The CUPS utilities (whose source code is included with the simulations with this book), include additional object-oriented programs for a complete graphical user interface, including pull-down menus, sliders, buttons, hotkeys, and mouse clicking and dragging. They also include routines for creating contour, two-dimensional (2-D) and 3-D plots, and a function parser. The CUPS utilities have been provided in source code form to enable users to run the simulations under future generations of Borland/Turbo Pascal. If you do run under future generations of Turbo or Borland Pascal on the PC, the utilities and programs will need to be recompiled. You will also need to create a new egavga.obj file which gets combined with the programs when an executable version is created—thereby avoiding the need to have separate (egavga.bgi) driver files. These CUPS utilities are also available to users who wish to use them for their own projects.

One element not included in the utilities is a procedure for creating hard copy based on screen images. When hard copy is desired, those PC users with the appropriate graphics driver (graphics.com), may be able to produce high-quality screen images by depressing the **PrintScreen** key. If you do not have the graphics software installed to get screen dumps, select **Configuration | Print Screen**, and follow the directions. Moreover, public domain software also exists for capturing screen images, and for producing PostScript files, but the user should be aware that such files are often quite large, sometimes over 1 MB, and they require a PostScript printer driver to produce.

One feature of the CUPS utilities that can improve the quality of hard copy produced from screen captures is a procedure for switching colors. This capability is important because the gray scale rendering of colors on black-and-white printers may create poor contrasts if the original (default) color assignments are used. To access the CUPS utility for changing colors, the user need only choose **Configuration** under the **File** menu when the program is first initiated, or at any later time. Once you have chosen **Configuration**, to change colors you need to click the mouse on the **Change Colors** bar, and you will be presented with a 16 by 16 matrix of radio buttons that will allow you to change any color to any other color, or else to use predefined color switches, such as a color "reversal," or a conversion of all light colors to black, and all dark colors to white. (The screen captures given in this book were produced using the "reverse" color map.) Any such color changes must be redone when the program is restarted.

Other system parameters may likewise be set from the **File | Configuration** menu item. These include the path for temporary files that the program may create (or want to read), the mouse "double click" speed—important for those with slow reflexes—an added time delay to slow down programs on computers that are too fast, and a "check memory" option—primarily of interest to those making program modifications.

Those users wishing more information on the CUPS utilities should consult the CUPS Utilities Manual, written by Jaroslaw Tuszynski and William MacDonald, published by John Wiley and Sons. However, it is not necessary for casual users of CUPS programs to become familiar with the utilities. Such familiarity would only be important to someone wishing to write their own simulations using the utilities. The utilities are freely available for this purpose, for unrestricted noncommercial production and distribution of programs. However, users of the utilities who wish to write programs for commercial distribution should contact John Wiley and Sons.

1.5 Communicating With the Authors

Users of these programs should not expect that run-time errors will never occur! In most cases, such run-time errors may require only that the user restart the program; but in other cases, it may be necessary to reboot the computer, or even turn it off and on. The causes of such run-time errors are highly varied. In some cases, the program may be telling you something important about the physics or the numerical method. For example, you may be trying to use a numerical method beyond its range of applicability. Other types of run-time errors may have to do with memory or other limitations of your computer. Finally, although the programs in this book have been extensively tested, we cannot rule out the possibility that they may contain errors. (Please let us know if you find any! It would be most helpful if such problems were communicated by electronic mail, and with complete specificity as to the circumstances under which they arise.)

It would be best if you communicated such problems directly to the author of each program, and simultaneously to the editors of this book (the CUPS Directors), via electronic mail—see addresses listed below. Please feel free to communicate any suggestions about the programs and text which may lead to improvements in future editions. Since the programs have been provided in source code form, it will be possible for you to make corrections of any errors that you or we find in the future—provided that you send in the registration card at the back of the book, so that you can be notified. The fact that you have the source code will also allow you to make modifications and extensions of the programs. We can assume no responsibility for errors that arise in programs that you have modified. In fact, we strongly urge you to change the program name, and to add a documentary note at the beginning of the code of any modified programs that alerts other potential users of any such changes.

1.6 CUPS Courses and Developers

- **CUPS Directors**
  Maria Dworzecka, George Mason University (cups@gmuvax.gmu.edu)
  Robert Ehrlich, George Mason University (cups@gmuvax.gmu.edu)
  William MacDonald, University of Maryland (w_macdonald@umail.umd.edu)

- **Astrophysics**
  J. M. Anthony Danby, North Carolina State University (n38hs901@ncuvm.ncsu.edu)
  Richard Kouzes, Battelle Pacific Northwest Laboratory (rt_kouzes@pnl.gov)
  Charles Whitney, Harvard University (whitney@cfa.harvard.edu)

- **Classical Mechanics**
  Bruce Hawkins, Smith College (bhawkins@smith.edu)
  Randall Jones, Loyola College (rsj@loyvax.bitnet)

- **Electricity and Magnetism**
  Robert Ehrlich, George Mason University (rehrlich@gmuvax.gmu.edu)
  Lyle Roelofs, Haverford College (lroelofs@haverford.edu)
  Ronald Stoner, Bowling Green University (stoner@andy.bgsu.edu)
  Jaroslaw Tuszynski, George Mason University (cups@gmuvax.gmu.edu)

- **Modern Physics**
  Douglas Brandt, Eastern Illinois University (cfdeb@ux1.cts.eiu.edu)
  John Hiller, University of Minnesota, Duluth (jhiller@d.umn.edu)
  Michael Moloney, Rose Hulman Institute (moloney@nextwork.rose-hulman.edu)

- **Nuclear and Particle Physics**
  Roberta Bigelow, Willamette University (rbigelow@willamette.edu)
  John Philpott, Florida State University (philpott@fsunuc.physics.fsu.edu)
  Joseph Rothberg, University of Washington (rothberg@phast.phys.washington.edu)

- **Quantum Mechanics**
  John Hiller, University of Minnesota, Duluth (jhiller@d.umn.edu)
  Ian Johnston, University of Sydney (idj@suphys.physics.su.oz.au)
  Daniel Styer, Oberlin College (dstyer@physics.oberlin.edu)

- **Solid State Physics**
  Graham Keeler, University of Salford (g.j.keeler@sysb.salford.ac.uk)
  Roger Rollins, Ohio University (rollins@chaos.phy.ohiou.edu)
  Steven Spicklemire, University of Indianapolis (steves@truevision.com)

- **Thermal and Statistical Physics**
  Harvey Gould, Clark University (hgould@vax.clarku.edu)
  Lynna Spornick, Johns Hopkins University
  Jan Tobochnik, Kalamazoo College (jant@kzoo.edu)

- **Waves and Optics**
  G. Andrew Antonelli, Wolfgang Christian, and Susan Fischer, Davidson College (wc@phyhost.davidson.edu)
  Robin Giles, Brandon University (giles@brandonu.ca)
  Brian James, Salford University (b.w.james@sysb.salford.ac.uk)

1.7 Descriptions of all CUPS Programs

Each of the computer simulations in this book (as well as those in the eight other books comprised by the CUPS Project) are described below. The individual headings under which programs appear correspond to the nine CUPS courses. In several cases, programs are listed under more than one course. The number of programs listed under the Astrophysics, Modern Physics, and Thermal Physics courses is appreciably greater than the others, because several authors have opted to subdivide their programs into many smaller programs. Detailed inquiries regarding CUPS programs should be sent to the program authors.

ASTROPHYSICS PROGRAMS

STELLAR (Stellar Models), written by Richard Kouzes, is a simulation of the structure of a static star in hydrodynamic equilibrium. This provides a model of a zero age main sequence star, and helps the user understand the physical processes that exist in stars, including how density, temperature, and luminosity depend on mass. Stars are self-gravitating masses of hot gas supported by thermodynamic processes fueled by nuclear fusion at their core. The model integrates the four differential equations governing the physics of the star to reach an equilibrium condition which depends only on the star's mass and composition.

EVOLVE (Stellar Evolution), written by Richard Kouzes, builds on the physics of a static star, and considers (1) how a gas cloud collapses to become a main sequence star, and (2) how a star evolves from the main sequence to its final demise. The model is based on the same physics as the STELLAR program. Starting from a diffuse cloud of gas, a protostar forms as the cloud collapses and reaches a sufficient density for fusion to begin. Once a star reaches equilibrium, it remains for most of its life on the main sequence, evolving off after it has consumed its fuel. The final stages of the star's life are marked by rapid and dramatic evolution.

BINARIES is the driver program for all Binaries programs (**VISUAL1, VISUAL2, ECLIPSE, SPECTRO, TIDAL, ROCHE, and ACCRDISK**).

VISUAL1 (Visual Binaries—Proper Motion), written by Anthony Danby, enables you to visualize the proper motion in the sky of the members of a visual binary system. You can enter the elements of the system and the mass ratio, as well as the speed at which the center of mass moves across the screen. The program also includes an animated three-dimensional demonstration of the elements.

VISUAL2 (Visual Binaries—True Orbit), written by Anthony Danby, enables you to select an apparent orbit for the secondary star with arbitrary eccentricity, with the primary at any interior point. The elements of the orbit are displayed. You can see the orbit animated in three dimensions, or can make up a set of "observations" based on the apparent orbit.

ECLIPSE (Eclipsing Binaries), written by Anthony Danby, shows simultaneously either the light curve and the orbital motion or the light curve and an animation of the eclipses. You can select the elements of the orbit and radii and magnitudes of the stars. A form of limb-darkening is also included as an option.

SPECTRO (Spectroscopic Binaries), written by Anthony Danby, allows you to select the orbital elements of a spectroscopic binary, and then shows simultaneously the velocity curve, the orbital motion, and a moving spectral line.

TIDAL (Tidal Distortion of a Binary), written by Anthony Danby, models the motion of a spherical secondary star around a primary that is tidally distorted by the secondary. You can select orbital elements, masses of the stars, a parameter describing the tidal lag, and the initial rate of rotation of the primary. The equations are integrated over a time interval that you specify. Then you can see the changes of the orbital elements, and the rotation of the primary, with time. You can follow the motion in detail around each revolution, or in a form where the equations have been averaged around each revolution.

ROCHE (The Photo-Gravitational Restricted Problem of Three Bodies), written by Anthony Danby, follows the two-dimensional motion of a particle that is subject to the gravitational attraction of two bodies in mutual circular orbits, and also, optionally, radiation pressure from these bodies. It is intended, in part, as background for the interpretation of the formation of accretion disks. Curves of zero velocity (that limit regions of possible motion) can be seen. The orbits can also be followed using Poincaré maps.

ACCRDISK (Formation of an Accretion Disk), written by Anthony Danby, follows some of the dynamical steps in this process. The dynamics is valid up to the initial formation of a hot spot, and qualititative afterward.

NBMENU is the driver program for all programs on the motion of N interacting bodies: **TWOGALAX, ASTROIDS, N-BODIES, PLANETS, PLAYBACK, and ELEMENTS**.

TWOGALAX (The Model of Wright and Toomres), written by Anthony Danby, is concerned with the interaction of two galaxies. Each consists of a central gravitationally attracting point, surrounded by rings of stars (which are attracted, but do not attract). Elements of the orbits of one galaxy relative to the other are selected, as is the initial distribution and population of the rings. The motion can be viewed as projected into the plane of the orbit of the galaxies, or simultaneously in that plane and perpendicular to it. The positions can be stored in a file for later viewing.

ASTROIDS (N-Body Application to the Asteroids), written by Anthony Danby, uses the same basic model, but a planet and a star take the place of the galaxies and the asteroids replace the stars. Emphasis is on asteroids all having the same period, with interest on periods having commensurability with the period of the planet. The orbital motion of the system can be followed. The positions can be stored in a file for later viewing. An asteroid can be selected, and the variation of its orbital elements can then be followed.

NBODIES (The Motion of N Attracting Bodies), written by Anthony Danby, allows you to choose the number of bodies (up to 20) and the total energy of the system. Initial conditions are chosen at random, consistent with this energy, and the resulting motion can be observed. During the motion various quantities, such as the kinetic energy, are displayed. The positions can be stored in a file for later viewing.

PLANETS (Make Your Own Solar System), written by Anthony Danby, is similar to the preceding program, but with the bodies interpreted as a star with planets. Initial conditions are specified through the choice of the initial elements of the planets. The positions can be stored in a file for later viewing.

PLAYBACK, written by Anthony Danby, enables a file stored by one of the preceding programs to be viewed.

ELEMENTS (Orbital Elements of a Planet), written by Anthony Danby, shows a three-dimensional animation that can be viewed from any angle.

GALAXIES is the driver program for Galactic Kinematics programs: **ROTATION, OORTCONS, and ARMS21CM**.

ROTATION (The Rotation Curve of a Galaxy), written by Anthony Danby, first prompts you to "design" a galaxy, consisting of a central mass and up to five spheroids (that can be visible or invisible). It then displays the galaxy and can show the animated rotation or the rotation curve.

OORTCONS (Galactic Kinematics and Oort's Constants), written by Anthony Danby, allows you to design your galaxy, choose the location of the "sun" and a local region around it, and the to observe the kinematics in this region. It also shows graphs of radial velocity and proper motion in comparison with the linear approximation, and computes the Oort constants.

ARMS21CM (The Spiral Structure of a Galaxy), written by Anthony Danby, allows you to design your galaxy, construct a set of spiral arms, and select the position of the "sun." Then, for different galactic longitudes, you can see observed profiles of 21 cm lines.

ATMOS (Stellar Atmospheres), written by Charles Whitney, permits the user to select a constellation, see it mapped on the computer screen, point to a star, and see it plotted on a brightness-color diagram. The user's task is to build a model atmosphere that imitates the photometric properties of observed stars. This is done by specifying numerical values for three basic stellar parameters: radius, mass, and luminosity. The program then builds the model and displays it on the brightness-color diagram, and it also plots the spectrum and the detailed thermodynamic structure of the atmosphere. With this program the user may investigate the relation between stellar parameters and the thermal properties of the gas in the atmosphere. Two atmospheres may be superposed on the graphs, for easier comparison.

PULSE (Stellar Pulsations), written by Charles Whitney, illustrates stellar pulsation by simulating the thermo-mechanical behavior of a "star" modeled by a self-gravitating gas divided by spherical elastic shells. The elastic shells resemble a set of coupled oscillators. The program solves for the modes of small-amplitude motion, and it uses Fourier synthesis to construct motions for arbitrary starting conditions. The screen displays the thermodynamic structure and surface properties, such as temperature, pressure, and velocity. Animation displays the nature of the pulsation. By showing the motions, temperatures, and energy flux, the program demonstrates the heat engine acting inside the pulsating star. The motions of the shells and the spatial Fourier decomposition into eigenmodes are displayed simultaneously, and this will help you visualize the meaning of the Fourier components.

CLASSICAL MECHANICS PROGRAMS

GENMOT (The Motion Generator), written by Randall Jones, allows you to solve numerically any differential equation of motion for a system with up to three degrees of freedom and display the time evolution of the system in a wide variety of formats. Any of the dynamical variables or any function of those variables may be displayed graphically and/or numerically and a wide range of animations may be constructed. Since the Motion Generator can be used to solve any second-order differential equation, it can also be used to study systems analyzed by Lagrangian methods. Real world coordinates may be constructed as functions of generalized coordinates so that simulations of the actual system can be constructed.

ROTATE (Rotation of 3-D Objects), written by Randall Jones, is designed to aid in the visualization of the dynamical variables of rotational motion. It will allow you to observe the 3-D motion of rotating objects in a controlled fashion, running the simulation faster, slower, or in reverse while displaying the corresponding evolution of the angular velocity, the angular momentum and the torque. It will display the motion from the fixed frame and from the body frame to help in understanding the translation between these two descriptions of the motion. By using the stereographic feature of the program you can create a genuine 3-D representation of the motion of the quantities.

COUPOSC (Coupled Oscillators), written by Randall Jones, is designed to investigate a wide range of harmonic systems. Given a set of objects and springs connected in one or two dimensions, the simulation can solve the problem by generating the normal mode frequencies and their corresponding motions. It can take any set of initial conditions and resolve them into their component normal mode motions or take any set of initial mode occupations and display the corresponding motions of the objects. It can also determine the motion of the system when it is acted on by external forces. In this case the total forces are no longer harmonic, so the solution is generated numerically. The harmonic analysis, however, still provides an important tool for investigating and understanding the subsequent motion.

ANHARM (Anharmonic Oscillators), written by Bruce Hawkins, simulates oscillations of various types: pendulum, simple harmonic oscillator, asymmetric, cubic, Vanderpol, and a mass in the center of a spring with fixed ends. Nonlinear behavior is emphasized. The user may choose to view one to four graphs of the motion simultaneously, along with the potential diagram and a picture of the moving object. Graphs that may be viewed are x vs. t, v vs. t, v vs. x, the Poincaré diagram, and the return map. Tools are provided to explore parameter space for regions of interest. Fourier analysis is available, resonance diagrams can be plotted, and the period can be plotted as a function of energy. Includes a tutorial demonstrating the usefulness of phase plots and Poincaré plots.

ORBITER (Gravitational Orbits), written by Bruce Hawkins, simulates the motion of up to five objects with mutually gravitational attraction, and any reasonable number of additional objects moving in the gravitation field of the first five. The motion may be viewed in up to six windows simultaneously: windows centered on a particular body, on the center of mass, stationary in the universe frame, or rotating with the line joining the two most massive bodies. A menu of available systems includes the solar system, the sun/earth/moon system; the sun, Jupiter, and its moons; the sun, earth, and Jupiter, demonstrating retrograde motion; the sun, Jupiter, and a comet; and a pair of binary stars with a comet. Bodies may be added to any system, or a new system created using either numerical coordinates or the mouse. Bodies may be replicated to demonstrate the sensitivity of orbits to initial conditions.

COLISION (Collisions), written by Bruce Hawkins, simulates two-body collisions under any of a number of force laws: Coulomb with and without shielding and truncation, hard sphere, soft sphere (harmonic), Yukawa, and Woods-Saxon. Collision may be viewed in the laboratory and center of mass systems simultaneously, with or without momentum diagrams. Includes a tutorial on the usefulness of the center of mass system, one on the kinematics of relativistic collisions, and one on cross section. Plots cross section against scattering parameter, and compares collisions at different parameters.

ELECTRICITY AND MAGNETISM PROGRAMS

FIELDS (Analysis of Vector and Scalar Fields), written by Jarek Tuszynski, displays scalar and vector fields for any algebraic or trigonometric expression entered by the user. It also computes numerically the divergence, curl, and Laplacian for the vector fields, and the gradient and Laplacian for the scalar fields. Simultaneous displays of selected quantities are provided in user-selected planes, using vector, contour, or 3-D plots. The program also allows the user to define paths along which line integrals are computed.

GAUSS (Gauss' Law), written by Jarek Tuszynski, treats continuous charge distributions having spherical or cylindrical symmetry, and those that vary as a function of the x-coordinate only. The program allows the user to enter an arbitrary function to define either the electric field magnitude, the potential, or the charge density. It then computes the other two functions by numerical differentiation or integration, and displays all three functions. Finally, the program allows the user to enter a "comparison function," which is plotted on the same graph, so as to check whether his analytic solutions are correct.

POISSON (Poisson's Equation Solved on a Grid), written by Jarek Tuszynski, solves Poisson's equation iteratively on a 2-D grid using the method of simultaneous over-relaxation. The user can draw arbitrary systems consisting of line charges, and charged conducting cylinders, plates, and wires, all infinite in extent perpendicular to the grid. After iteratively solving Poisson's equation, the program displays the results for the potential, electric field, or the charge density (found from the Laplacian of the potential), in the form of contour, vector, or 3-D plots. In addition, many other program features are available, including the ability to specify surfaces, along which the potential varies according to some algebraic function specified by the user.

IMAG&MUL (Image Charges and Multipole Expansion), written by Lyle Roelofs and Nathaniel Johnson, allows users to explore two approaches to the solution of Laplace's equation—the image charge method and expansion in multipole moments. In the image charge mode (IC) the user is presented with a variety of configurations involving conducting planes and point charges and is asked to "solve" each by placing image charges in the appropriate locations. The program displays the electric field due to all point charges, real and image, and a solution can be regarded as successful with the field due to all charges is everywhere orthogonal to all conducting surfaces. Solutions can then be examined with a variety of included software "tools." The multipole expansion (ME) mode of the program also permits a "hands-on" exploration of standard electrostatic problems, in this case the "exterior" problem, i.e., the determination of the field outside a specified equipotential surface. The program presents the user with a variety of azimuthally symmetric equipotential surfaces. The user "solves" for the full potential by adding chosen amounts of the (first six) multipole moments. The screen shows the contours of the summed potential and the problem is "solved" when the innermost contour matches the given equipotential surface as closely as possible.

ATOMPOL (Atomic Polarization), written by Lyle Roelofs and Nathaniel Johnson, is an exploration of the phenomenon of atomic polarization. Up to 36 atoms of controllable polarizability are immersed in an external electric field. The program solves for and displays the field throughout the region in which the atoms are located. A closeup window shows the polarization of selected atoms and software "tools" allow for further analysis of the resulting electric fields. Use of this program improves the student's understanding of polarization, the interaction of polarized entities, and the atomic origin of macroscopic polarization, the latter via study of closely spaced clusters of polarizable atoms.

DIELECT (Dielectric Materials), written by Lyle Roelofs and Nathaniel Johnson, is a simulation of the behavior of linear dielectric materials using a cell-based approach. The user controls either the polarization or the susceptibility of each cell in a (25 × 25) grid (with assumed uniformity in the third direction). Full self-consistent solutions are obtained via an iterative relaxation method and the fields P, E, or D are displayed. The student can investigate the self-interactions of polarized materials and many geometrical effects. Use of this program aids the student in developing understanding of the subtle relations among and meaning of P, E, and D.

ACCELQ (Fields From an Accelerated Charge), written by Ronald Stoner, simulates the electromagnetic fields in the plane of motion generated by a point charge that is moving and accelerating in two dimensions. The user chooses from among seven predefined trajectories, and sets the values of maximum speed and viewing time. The electric field pattern is recomputed after each change of trajectory or parameter; thereafter, the user can investigate the electric field, magnetic field, retarded potentials, and Poynting-vector field by using the mouse as a field probe, by using gridded overlays, or by generating plots of the various fields along cuts through the viewing plane.

QANIMATE (Fields From an Accelerated Charge—Animated Version), written by Ronald Stoner, is an interactive animation of the changing electric field pattern generated by a point electric charge moving in two dimensions. Charge motion can be manipulated by the user from the keyboard. The display can include electric field lines, radiation wave fronts, and their points of intersection. The motion of the charge is controlled by the using **arrow** keys to accelerate and steer much like the accelerator and steering wheel of a car, except that acceleration must be changed in increments, and the **Space** bar can used to engage or disengage the steering. With steering engaged, the charge will move in a circle. Unless the acceleration is made zero, the speed will increase (or decrease) to the maximum (minimum) possible value. At constant speed and turning rate, the charge can be controlled by the **Space** bar alone.

EMWAVE (Electromagnetic Waves), written by Ronald Stoner, uses animation to illustrate the behavior of electric and magnetic fields in a polarized plane electromagnetic wave. The user can choose to observe the wave in free space, or to see the effect on the wave of incidence on a material interface, or to see the effects of optical elements that change its polarization. The user can change the polarization state of the incident wave by specifying its Stokes parameters. Standing electromagnetic waves can be simulated by combining the incident traveling wave with a reflected wave of the same amplitude. The user can do that by choosing appropriate values of the physical properties of the medium on which the incident wave impinges in one of the animations.

MAGSTAT (Magnetostatics), written by Ronald Stoner, computes and displays magnetic fields in and near magnetized materials. The materials are uniform and have 3-D shapes that are solids of revolution about a vertical axis. The shape of the material can be modified or chosen from a data input screen. The user has the option of generating the fields produced by a permanently and uniformly magnetized object, or of generating the fields of a magnetizable object placed in an otherwise uniform external field. Besides choosing the shape and aspect ratio of the object, the user can vary the magnetic permeability of the magnetizable material, and choose among three fields to display: magnetic induction (B), magnetic field strength (H), and magnetization (M). Each of these fields can be displayed or explored in several different ways. The algorithm for computing the fields uses a superposition of Chebyschev polynomial approximants to the H field due to "rings" of "magnetic charge."

MODERN PHYSICS PROGRAMS

NUCLEAR (Nuclear Energetics and Nuclear Counting), written by Michael Moloney, deals with basic nuclear properties related to mass, charge, and energy, for approximately 1900 nuclides. Graphs are available involving binding energy, mass, and Q values of a variety of nuclear reactions, including alpha and beta decays. Part 2 deals with simulating the statistics of counting with a Geiger-Muller tube. This part also simulates neutron activation, and the counting behavior as neutron flux is turned on and off. Finally, a decay chain from A to B to C is simulated, where half-lives may be changed, and populations are graphed as a function of time.

GERMER (Davisson-Germer and G. P. Thomson Experiments), written by Michael Moloney, simulates both the Davisson-Germer and G. P. Thomson experiments with electrons scattering from crystalline materials. Stress is laid on the behavior of electrons as waves; similarities are noted with scattering of x-rays. The exercises encourage students to understand why peaks and valleys in scattered electrons occur where they do.

QUANTUM (one-dimensional Quantum Mechanics), written by Douglas Brandt, is a program that has four sections. The first section allows users to investigate the uncertainty principle for specified wavefunctions in position or momentum space. The second section allows users to investigate the time evolution of wavepackets under various dispersion relations. The third section allows users to investigate solutions to Schrödinger's equation for asymptotically free solutions. The user can input a barrier and the program calculates reflection and transmission coefficients for a range of energies and show wavepacket time evolution for the barrier potential. The fourth section is similar to the third, except that it allows the user to investigate bound solutions to Schrödinger's equation. The program calculates the bound state Hamiltonian eigenvalues and spatial eigenfunctions.

RUTHERFD (Rutherford Scattering), written by Douglas Brandt, is a program for investigating classical scattering of particles. A scattering potential can be chosen from a list of predefined potentials or an arbitrary potential can be input by the user. The computer generates scattering events by randomly picking impact parameters from a distribution defined by beam parameters specified by the user. It displays the results of the scattering on a polar histogram and on a detailed histogram to help users gain insight into differential scattering cross section. A scintillation mode can be chosen for users that want more appreciation of the actual experiments of Geiger and Marsden. A "guess the scatterer" mode is available for trying to gain appreciation of how scattering experiments are used to infer properties of the scatterers.

SPECREL (Special Relativity), written by Douglas Brandt, is a program to investigate special relativity. The first section is to investigate change of coordinate systems through Minkowski diagrams. The user can define coordinates of objects in one reference frame and the computer calculates the coordinates in a user-selectable coordinate system and displays the objects in both reference frames. The second section allows users to view clocks that are in relative motion. A clock can be given an arbitrary trajectory through space-time and the readings of various clocks can be viewed as the clock follows that trajectory. A third section allows users to observe collisions in different reference frames that are related by Lorentz transformations or by Gallilean transformations.

LASER (Lasers), written by Michael Moloney, simulates a three-level laser, with the user in control of energy level parameters, temperature, pump power, and end mirror transmission. Atomic populations may be graphically tracked from thermal equilibrium through the lasing threshold. A mirror cavity simulation is available which uses ray tracing. This permits study of cavity stability as a function of mirror shape and position, as well as beam shape characteristics within the cavity.

HATOM (Hydrogenic Atoms), written by John Hiller, computes eigenfunctions and eigenenergies for hydrogen, hydrogenic atoms, and single-electron diatomic ions. Hydrogenic atoms may be exposed to uniform electric and magnetic fields. Spin interactions are not included. The magnetic interaction used is the quadratic Zeeman term; in the absence of spin-orbit coupling, the linear term adds only a trivial energy shift. The unperturbed hydrogenic eigenfunctions are computed directly from the known solutions. When external fields are included, approximate results are obtained from basis-function expansions or from Lanczos diagonalization. In the diatomic case, an effective nuclear potential is recorded for use in calculation of the nuclear binding energy.

NUCLEAR AND PARTICLE PHYSICS PROGRAMS

NUCLEAR (Nuclear Energetics and Counting), written by Michael Moloney, is included here, but is described under the Modern Physics heading.

SHELLMOD (Nuclear Models), written by Roberta Bigelow, calculates energy levels for spherical and deformed nuclei using the single particle shell model. You can explore how the nuclear potential shape, the spin-orbit interaction, and deformation affect both the order and spacing of nuclear energy levels. In addition, you will learn how to predict spin and parity for single particle states.

NUCRAD (Interaction of Radiation With Matter), written by Roberta Bigelow, is a simulation of alpha particles, muons, electrons, or photons interacting with matter. You will develop an understanding of how ranges, energy losses, and random particle paths depend on materials, radiation, and incident energy. As a specific application, you can explore photon and electron interactions in a sodium iodide crystal which determines the energy response of a radiation detector.

ELSCATT (Electron-Nucleus Scattering), by John Philpott, is an interactive software tool that demonstrates various aspects of electron scattering from nuclei. Specific features include the relativistic kinematics of electron scattering, densities and form factors for elastic and inelastic scattering, and the nuclear Coulomb response. The simulation illustrates how detailed nuclear structure information can be obtained from electron scattering measurements.

TWOBODY (Two-Nucleon Interactions), by John Philpott, is an interactive software tool that illuminates many features of the two-nucleon problem. Bound state wavefunctions and properties can be calculated for a variety of interactions that may include non-central parts. Phase shifts and cross sections for pp, pn, and nn scattering can be calculated and compared with those obtained experimentally. Spin-polarization features of the cross sections can be extensively investigated. The simulation demonstrates the richness of the two-nucleon data and its relation to the underlying nucleon-nucleon interaction.

RELKIN (Relativistic Kinematics), by Joseph Rothberg, is an interactive program to permit you to explore the relativistic kinematics of scattering reactions and two-body particle decays. You may choose from among a large number of initial and final states. The initial momentum of the beam particle and the center of mass angle of a secondary can also be specified. The program displays the final state vector momenta in both the lab system and center of mass system along with numerical values of the most important kinematic quantities. The program may be run in a Monte Carlo mode, displaying a scatter plot and histogram of selected variables. The particle data base may be modified by the user and additional reactions and decay modes may be added.

DETSIM (Particle Detector Simulation), by Joseph Rothberg, is an interactive tool to allow you to explore methods of determining parameters of a decaying particle or scattering reaction. The program simulates the response of high-energy particle detectors to the final-state particles from scattering or decays. The detector size and location may be specified by the user as well as its energy and spatial resolution. If the program is run in a Monte Carlo mode, detector hit information for each event is written to a file. This file can be read by a small reconstruction and plotting program. You can easily modify one of the example reconstruction programs that are provided to determine the mass, momentum, and other properties of the initial particle or state.

QUANTUM MECHANICS PROGRAMS

BOUND1D (Bound States in One Dimension), written by Ian Johnston, is a tool which allows you to explore energy eigenfunctions for an electron in various potential wells, which can be square, parabolic, ramped, asymmetric, double, or Coulombic. The first part of the program deals with finding the eigenvalues and eigenfunctions of different wells. You may find them yourself, using a "hunt and shoot" method, or else the program will compute the eigenvalues automatically, by counting the number of nodes to determine where the eigenvalues occur. The second part of the program looks at properties of eigenfunctions normalization, orthogonality, and the evaluation of many kinds of overlap integrals. The third part examines the time development of general states made up of a superposition of bound state eigenfunctions. Facility is provided for you to incorporate your own procedures to specify different potential wells or different overlap integrals.

SCATTR1D (Scattering in One Dimension), written by John Hiller, solves the time-independent Schrödinger equation for stationary scattering states in one-dimensional potentials. The wavefunction is displayed in a variety of ways, and the transmission and reflection probabilities are computed. The probabilities may be displayed as functions of energy. The computations are done by numerically integrating the Schrödinger equation from the region of the transmitted wave, where the wavefunction is known up to some overall normalization and phase, to the region of the incident wave. There the reflected and incident waves are separated. The potential is assumed to be zero in the incident region and constant in the transmitted region.

QMTIME (Quantum Mechanical Time Development), written by Daniel Styer, simulates quantal time development in one dimension. A variety of initial wave packets (Gaussian, Lorentzian, etc.) can evolve in time under the influence of a variety of potential energy functions (step, ramp, square well, harmonic oscillator, etc.) with or without an external driving force. A novel visualization technique simultaneously displays the magnitude and phase of complex-valued wave functions. Either position-space or momentum-space wave functions, or both, can be shown. The program is particularly effective in demonstrating the classical limit of quantum mechanics.

LATCE1D (Wavefunctions on a one-dimensional Lattice), written by Ian Johnston, is a tool which allows you to explore energy eigenfunctions for an electron in a lattice made up of a number of simple potential wells (up to twelve), which can be square, parabolic, or Coulombic. You may find the eigenvalues yourself, using a "hunt and shoot" method, or allow the program to compute them automatically. You can firstly explore regular lattices, where all wells are the same and spaced at regular intervals. These will demonstrate many of the properties of regular crystals, particularly the existence of energy bands. Secondly you can change the width, depth or spacing of any of the wells, which will mimic the effect of impurities or other irregularities in a crystal. Lastly you can apply an external electric across the lattice. Facility is provided for you to incorporate your own procedures to calculate wells, lattice arrangements or external fields of their own choosing.

BOUND3D (Bound States in Three Dimensions), written by Ian Johnston, is a tool which allows you to explore energy eigenfunctions for a particle in a spherically symmetric potential well, which can be square, parabolic, Coulombic, or several other shapes of importance in molecular or nuclear applications. The first part of the program deals with finding the eigenvalues and eigenfunctions of different wells, assuming that the angular part of the wavefunctions are spherical harmonics. You may find them yourself for a given angular momentum quantum number using a "hunt and shoot" method, or else the program will compute the eigenvalues automatically, by counting the number of nodes to determine where the eigenvalues occur. The second part of the program looks at properties of eigenfunctions normalization, orthogonality, and the evaluation of many kinds of overlap integrals. Facility is provided for you to incorporate your own procedures to specify different potential wells or different overlap integrals.

IDENT (Identical Particles in Quantum Mechanics), written by Daniel Styer, shows the probability density associated with the symmetrized, antisymmetrized, or nonsymmetrized wave functions of two noninteracting particles moving in a one-dimensional infinite square well. It is particularly valuable for demonstrating the effective interaction of noninteracting identical particles due to interchange symmetry requirements.

SCATTR3D (Scattering in Three Dimensions), written by John Hiller, performs a partial-wave analysis of scattering from a spherically symmetric potential. Radial and 3-D wavefunctions are displayed, as are phase shifts, and differential and total cross sections. The analysis employs an expansion in the natural angular momentum basis for the scattering wavefunction. The radial wavefunctions are computed numerically; outside the region where the potential is important they reduce to a linear combination of Bessel functions which asymptotically differs from the free radial wavefunction by only a phase. Knowledge of these phase shifts for the dominant values of angular momentum is used to approximate the cross sections.

CYLSYM (Cylindrically Symmetric Potentials), written by John Hiller, solves the time-independent Schrödinger equation Hu=Eu in the case of a cylindrically symmetric potential for the lowest state of a chosen parity and magnetic quantum number. The method of solution is based on evolution in imaginary time, which converges to the state of the lowest energy that has the symmetry of the initial guess. The Alternating Direction Implicit method is used to solve a diffusion equation given by $HU = -\hbar\partial U/\partial t$, where H is the Hamiltonian that appears in the Schrödinger equation. At large times, U is nearly proportional to the lowest eigenfunction of H, and the expectation value $\langle H\rangle = \langle U \mid H \mid U\rangle/\langle U \mid U\rangle$ is an estimate for the associated eigenenergy.

SOLID STATE PHYSICS

LATCE1D (Wavefunctions for a one-dimensional Lattice), written by Ian Johnston, and included here, is described under the Quantum Mechanics heading.

SOLIDLAB (Build Your Own Solid State Devices), written by Steven Spicklemire, is a simulation of a semiconductor device. The device can be "drawn" by the user, and the characteristics of the device adjusted by the user during the simulation. The user can see how charge density, current density, and electric potential vary throughout the device during its operation.

LCAOWORK (Wavefunctions in the LCAO Approximation), written by Steven Spicklemire, is a simulation of the interaction of 2-D atoms within small atomic clusters. The atoms can be adjusted and moved around while their quantum mechanical wavefunctions are calculated in real time. The student can investigate the dependence of various properties of these atomic clusters on the properties of individual atoms, and the geometric arrangement of the atoms within the cluster.

PHONON (Phonons and Density of States), written by Graham Keeler, calculates and displays phonon dispersion curves and the density of states for a number of different 3-D cubic crystal structures. The displays of the dispersion curves show realistic curves and allow the user to study the effect of changing the interatomic forces between nearest and further neighbor atoms and, for diatomic crystal structures, changing the ratio of the atomic masses. The density of states calculation shows how the complex shapes of real densities of states are built up from simpler distributions for each mode of polarization, and enables the user to match the features of the distribution to corresponding features on the dispersion curves. In order to help with visualization of the crystal lattices involved, the program also shows 3-D projections of the different crystal structures.

SPHEAT (Calculation of Specific Heat), written by Graham Keeler, calculates and displays the temperature variation of the lattice specific heat for a number of different theoretical models, including the Einstein model and the Debye model. It also makes the calculation for a computer simulation of a realistic density of states, in which the user can vary the important parameters of the crystal, including those affecting the density of states. The program can display the results for a small region near the origin, and as a T-cubed plot to enable the user to investigate the low temperature limit of the specific heat, or in the form of the equivalent Debye temperature to enhance a study of the deviations from the Debye model. The Schottky specific heat anomaly can also be investigated.

BANDS (Energy Bands), written by Roger Rollins, calculates and displays, for easy comparison, the energy dispersion curves and corresponding wavefunctions for an electron in a 1-D symmetric $V(x) = V(-x)$ periodic potential of arbitrary shape and of strength V_0. The method used is based on an exact, non-perturbative approach so that the energy dispersion curves and band gaps can be obtained for large V_0. Wavefunctions can be displayed, and compared with one another, by clicking the mouse on the desired states on the energy dispersion curve. Changes in band structure can be followed as changes are made in the shape of the potential. The variation of the band gaps with V_0 is calculated and compared with the two opposite limits of very weak V_0 (perturbation method) and very strong V_0 (isolated atom). Even the experienced condensed matter researcher may be surprised by some of the results! Open-ended class discussions can result from the interesting physics found in these conceptually simple model calculations.

PACKET (Electron Wavepacket in a 1-D Lattice), written by Roger Rollins, shows a live animation, calculated in real time, demonstrating how an electron wavepacket in a metal or semiconducting crystal moves under the influence of external forces. The time-dependent Schrödinger equation is solved in a tight binding approximation, including the external force terms, and the motion of the wavepacket is obtained directly. The main objective of the simulation is to show that an electron wavepacket formed from states with energies near the top of an energy band is accelerated in a direction *opposite* to the direction of the external force; it has a *negative* effective mass! The simulation deals with motion in a 1-D lattice but the concepts are applicable to the full 3-D motion of an electron in a real crystal. Numerical experiments on the motion of the packet explore interesting physics questions such as: how does constant applied force affect the periodic motion of a packet? when does the usual semiclassical model fail? what happens to the dynamics of the packet when placed in a superlattice with lattice constant twice that of the original lattice?

THERMAL AND STATISTICAL PHYSICS PROGRAMS

ENGDRV, written by Lynna Spornick, is a driver program for **ENGINE, DIESEL, OTTO, and WANKEL**. These programs provide an introduction to the thermodynamics of engines.

ENGINE (Design Your Own Engine), written by Lynna Spornick, lets the user design an engine by specifying the processes (adiabatic, isobaric, isochoric [constant volume], and isothermic) in the engine's cycle, the engine type (reversible or irreversible), and the gas type (helium, argon, nitrogen, or steam). The thermodynamic properties (heat exchanged, work done, and change in internal energy) for each process and the engine's efficiency are computed.

DIESEL, OTTO, and WANKEL, written by Lynna Spornick, provide animations of each of these types of engine. Plots of the temperature versus entropy and the pressure versus volume for the cycles are shown with the engine's current thermodynamic conditions indicated.

PROBDRV, written by Lynna Spornick, is a driver program for **GALTON, POISEXP, TWOD, KAC, and STADIUM**. Subprograms GALTON, POISEXP, and TWOD provide an introduction to probability and subprograms KAC and STADIUM provide an introduction to statistics.

GALTON (A Galton Board), written by Lynna Spornick, models either a traditional Galton Board or a customized Galton Board with traps, reflecting, and/or absorbing walls. GALTON demonstrates the binominal and normal distributions, the laws of probability, and the central limit theorem.

POISEXP (Poisson Probability Distribution in Nuclear Decay), written by Lynna Spornick, uses the decay of radioactive atoms to describe the Poisson and the exponential distributions.

TWOD (2-D Random Walk), written by Lynna Spornick, models a random walk in two dimensions. A "drunk," taking equal-length steps, is required to walk either on a grid or on a plane. TWOD demonstrates the joint probability of two independent processes, the binominal distribution, and the Rayleigh distribution.

KAC (A Kac Ring), written by Lynna Spornick, uses a Kac ring to demonstrate that large mechanical systems, whose equations of motion are solvable and which obey time reversal and have a Poincaré cycle, can also be described by statistical models.

STADIUM (The Stadium Model), written by Lynna Spornick, uses a stadium model to demonstrate that there exist mechanical systems whose equations of motion are solvable but whose motion is not predictable because of the system's chaotic nature.

ISING (Ising Model in One and Two Dimensions), written by Harvey Gould, allows the user to explore the static and dynamic properties of the 1- and 2-D Ising model using four different Monte Carlo algorithms and three different ensembles. The choice of the Metropolis algorithm allows the user to study the Ising model at constant temperature and external magnetic field. The orientation of the spins is shown on the screen as well as the evolution of the total energy or magnetization. The mean energy, magnetization, heat capacity, and susceptibility are monitored as a function of the number of configurations that are sampled. Other computed quantities include the equilibrium-averaged energy and magnetization autocorrelation functions and the energy histogram. Important physical concepts that can be studied with the aid of the program include the Boltzmann probability, the qualitative behavior of systems near critical points, critical exponents, the renormalization group, and critical slowing down. Other algorithms that can be chosen by the user correspond to spin exchange dynamics (constant magnetization), constant energy (the demon algorithm), and single cluster Wolff dynamics. The latter is particularly useful for generating equilibrium configurations at the critical point.

MANYPART (Many Particle Molecular Dynamics), written by Harvey Gould, allows the user to simulate a dense gas, liquid, or solid in two dimensions using either molecular dynamics (constant energy, constant volume) or Monte Carlo (constant temperature, constant volume) methods. Both hard disks and the Lennard-Jones interaction can be chosen. The trajectories of the particles are shown as the system evolves. Physical quantities of interest that are monitored include the pressure, temperature, heat capacity, mean square displacement, distribution of the speeds and velocities, and the pair correlation function. Important physical concepts that can be studied with the aid of the program include the Maxwell-Boltzmann probability distribution, fluctuations, equation of state, correlations, and the importance of chaotic mixing.

FLUIDS (Thermodynamics of Fluids), written by Jan Tobochnik, allows the user to explore the fluid (gas and liquid) phase diagrams for the van der Waals model and water. The user chooses four phase diagrams from among the following choices: PT, Pv, vT, uT, sT, uv, and sv, where P is the pressure, T is the temperature, v is the specific volume, S is the specific entropy, and u is the specific internal energy. The program reads in the coexistence table for the van der Waals model and water, and uses it along with an empirical formula for the water free energy and the free energy derived from the van der Waals model. Given v and u, any thermodynamic quantity can be calculated. For the van der waals model thermodynamic quantities also can be calculated from the other thermodynamic state variables. The user can draw a straight line path in one phase diagram and see how this path looks in the other phase diagrams. The user also can extract all important thermodynamic data at any point in a phase diagram.

QMGAS1 (Quantum Mechanical Gas—Part 1), written by Jan Tobochnik, does the numerical calculations necessary to solve for the thermodynamic properties of quantum ideal gases, including photons in blackbody radiation, ideal bosons, phonons in the Debye theory, non-interacting fermions, and the classical limits of these systems. The user chooses the type of statistics (Bose-Einstein, Fermi-Dirac, or Maxwell-Boltzmann), the dimension of space, the form of the dispersion relation (restricted to simple powers), whether or not the particles have a non-zero chemical potential, and whether or not there is a Debye cutoff. The program then allows the user to build up a table of thermodynamic data, including the energy, specific heat, and chemical potential as a function of temperature. This data and various distribution functions and the density of states can be plotted.

QMGAS2 (Quantum Mechanical Gas—Part2), written by Jan Tobochnik, implements a Monte Carlo simulation of a finite number of quantum particles fluctuating between various states in a finite k-space (k is the wavevector). The program orders the possible energy states into an energy level diagram and then allows particles to move from one state to another according to the usual Boltzman probability distribution. Bosons are restricted so that they may not pass through each other on the energy level diagram; fermions are further restricted so that no two fermions may be in the same state; classical particles have no restrictions. In this way indistinguishability is correctly implemented for bosons and fermions. The user chooses the type of particle, the number of particles, the size and dimension of k-space, and the temperature. During the simulation the user sees a representation of the state occupancy and plots of the average energy, the instantaneous energy, and the distribution of energy among the states, also shown are results for the average energy, specific heat, and the occupancy of the ground state.

WAVES AND OPTICS PROGRAMS

DIFFRACT (Interference and Diffraction), by Robin Giles, simulates some of the fundamental wave behaviors in Fresnel and Fraunhofer Diffraction, and other Interference and Coherence effects. In particular you will be able to study diffraction phenomena associated with a point or a set of points and a slit or set of slits using the Huyghens construction. You can also use a method developed by Cornu—the Cornu Spiral—to examine diffraction from one or two slits or one or two obstacles. You can study Fresnel and Fraunhofer diffraction with a single slit or set of slits, a rectangular aperture and a circular aperture. Finally you can study Partial Coherence and fringe visibility in interference and diffraction observations. In the latter example you will be able to study the Michelson Stellar Interferometer and measure the separation distance in a double star and measure the diameter of single stars.

SPECTRUM (Applications of Interfence and Diffraction), by Robin Giles, simulates the uses and modes of operation of four important optical instruments—the Diffraction Grating, the Prism Spectrometer, the Michelson Interferometer and the Fabry-Perot Interferometer. You will look at the nature of the spectra, simulated interference patterns, and the question of resolving power.

WAVE (One-Dimensional Waves), by Wolfgang Christian, Andrew Antonelli, and Susan Fischer, uses finite difference methods to study the time evolution of the following partial differential equations: classical wave, Schrödinger, diffusion, Klein-Gordon, sine-Gordon, phi four, and double sine-Gordon. The user may vary the initial function and boundary conditions. Unique features of the program include mouse-driven drawing tools that enable the user to create sources, segments, and detectors anywhere inside the medium. Double-clicking on a segment allows the user to edit properties such as index of refraction or potential in order to model barrier problems such as thin film interference filters or the Ramsauer-Townsend effect in optics and quantum mechanics, respectively. Various types of analysis can be performed, including detector value, space-time, Fourier analysis and energy density.

CHAIN (One-Dimensional Lattice of Coupled Oscillators), by Wolfgang Christian, Andrew Antonelli, and Susan Fischer, allows the user to examine the time evolution of a 1-D lattice of coupled oscillators. These oscillators are represented on screen as a chain of masses undergoing vertical displacement. The program allows the user to examine how the application of Newtonian mechanics to these masses leads to traveling and standing waves. The relationship between the lattice spacing and other properties such as dispersion, band gaps, and cut-off frequency can be examined. Each mass can be assigned linear, quadratic, and cubic nearest neighbor interactions as well as a time-dependent external force function. Global properties such as the total energy in the lattice or the Fourier transform of the lattice can be displayed as well as the time evolution of a single mass's dynamical variables.

FOURIER (Fourier Analysis and Synthesis), written by Brian James, allows investigation of Fourier analysis and 1-D and 2-D Fourier transforms. In Fourier analysis users can choose from several predefined functions or enter their own functions either algebraically, numerically, or graphically. The build-up of a periodic function is illustrated as successive terms of the Fourier series are added in, and the effects of dispersion and attenuation on the evolution of the synthesized waveform can then be investigated. One- and two-dimensional discrete Fourier transforms can be produced for a range of standard and user-entered functions. The effects of filters on the inverse transforms are illustrated. The 2-D transforms are shown as surface and contour plots. Image processing can be illustrated by filtering the transforms of gray level images so that when the inverse transforms are displayed it can be seen that the images have been modified.

RAYTRACE (Ray Tracing and Lenses), by Brian James, lets the user explore the applications of ray tracing in geometrical optics. The fundamental principle of Fermat can be illustrated by plotting the path of a ray through two different materials between fixed points. The variation of the path of a ray through a region of changing refractive index can be used to investigate the formation of mirages. The variation of pulse delay in a fiber can be calculated as a function of its parameters and the characteristics of optical communication fibers are considered. The formation of primary and secondary rainbows due to dispersion of refractive index can be displayed. The matrix method of tracing rays through lenses can be used to investigate the images formed and show how aberrations in images arise and may be reduced.

QUICKRAY (Quick Ray Tracing), by John Philpott, can be used to demonstrate ray diagrams for a single thin lens or spherical mirror. The object and image are shown, along with the three principal rays that proceed from the object towards the observer. You can use the mouse to move the object, the position of the lens or mirror or to change the focal length of the lens or mirror. The principal rays are continously redrawn while any of these adjustments are made. The simulation handles converging and diverging lenses and concave and convex mirrors. Thus students can quickly get an intuitive feel for real and virtual image formation under a variety of circumstances.

Acknowledgments

The CUPS Project was funded by the National Science Foundation (under grant number PHY-9014548), and it has received support from the IBM Corporation, the Apple Corporation, and George Mason University.

2

The Motion Generator

Randall S. Jones

2.1 Introduction

To obtain a quick introduction to this simulation see the GENMOT walk-through in Appendix A.

Problems in dynamics usually involve two steps: first, the applied forces must be analyzed to determine the differential equations of motion describing the system; second, the differential equations must be solved to generate the subsequent motion. Sometimes the resulting differential equations can be solved in closed form, but more often, a numerical approach must be used.

This simulation allows you to solve numerically any differential equation of motion for a system with up to three degrees of freedom and display the time evolution of the system in a wide variety of formats. Any of the dynamical variables or any function of those variables may be displayed graphically and/or numerically and a wide range of animations may be constructed. Since the Motion Generator can be used to solve any second-order differential equation, it can also be used to study systems analyzed by Lagrangian methods. Spatial coordinates can be constructed as functions of generalized coordinates so that simulations of the actual system can be constructed.

To achieve this level of functionality, the Motion Generator requires a minimal amount of programming in Pascal; the functions describing the differential equation are entered as a procedure, another procedure is used to specify any functions of the dynamical variables that may be of interest, and two other procedures may be modified in order to create special animation effects. The programming required is very simple, but the advantage lies in the increased flexibility given the user for creating his or her own simulations.

A few example programs are provided to illustrate the capabilities of the program, but the real usefulness of the simulation is in the investigation of new systems. Careful guidance to setting up your own simulations is given in section 2.3.2 and 2.3.3.

2.2 Equations of Motion—Theory

There are many physical situations for which we know the nature of the forces acting on a particle and we know the initial position and initial velocity of the particle. The task is then to determine the subsequent motion; that is, to determine $x(t)$ (and hence $v(t)$). For many simple cases it is possible to determine $x(t)$ analytically (i.e., in terms of functions of t), but for real world problems, we usually must resort to some sort of numerical method. The simplest such approach is known as *Euler's method*, and most other methods are essentially improvements on this method.

2.2.1 Euler's Method

To develop the method we consider the motion of a single particle of mass m moving in one dimension. We assume that the force that acts on the particle is known. That force may depend on the time, on the position of the particle, and/or on the velocity of the particle so that we may write

$$Force = F(t, x, v)\,. \tag{2.1}$$

Note that it is also possible that the force depends on the positions and velocities of other particles, but we will ignore such cases for now. Later, it will be a simple matter to take other particles into account by extending this analysis to more than one particle.

Newton's second law and the definitions of acceleration and velocity are all that we need to develop the method:

$$\begin{aligned} F &= ma \\ a &= \lim_{\Delta t \to 0} \left(\frac{\Delta v}{\Delta t}\right) \\ v &= \lim_{\Delta t \to 0} \left(\frac{\Delta x}{\Delta t}\right). \end{aligned} \tag{2.2}$$

If we consider small but non-zero Δt, we can write

$$\begin{aligned} \Delta x &\approx v\Delta t \\ \Delta v &\approx a\Delta t\,. \end{aligned} \tag{2.3}$$

The first expression is fairly straightforward: if Δt is sufficiently small, the distance traveled by a particle in that time is given by the standard formula $d = vt$. Even if the velocity is changing as a function of time, for small enough Δt the velocity will not change "very much." The second equation has the same motivation: if Δt is sufficiently small, the acceleration will be essentially constant over that time interval.

Since we know the position and velocity at the initial time t_0,

$$\begin{aligned} x_0 &= x(t_0) \\ v_0 &= v(t_0)\,, \end{aligned} \tag{2.4}$$

we also know the force and hence the acceleration that acts on the object at that time:

$$F_0 = F(t_0, x_0, v_0) \rightarrow a_0 = F_0/m\,. \tag{2.5}$$

With this information, we can calculate the position and velocity of the object after a time Δt (i.e., at $t_1 = t_0 + \Delta t$) by using Eq. 2.3:

$$\begin{aligned} x_1 &= x_0 + v_0\Delta t \\ v_1 &= v_0 + a_0\Delta t\,. \end{aligned} \tag{2.6}$$

Once again, if Δt is sufficiently small, x_1 and v_1 will be very close to their correct values. Now, however, we can repeat the process starting with t_1, x_1, and v_1 to find the position and velocity of the particle at $t_2 = t_0 + 2\Delta t$:

$$\begin{aligned} a_1 &= F(t_1, x_1, v_1)/m \\ x_2 &= x_1 + v_1\Delta t \\ v_2 &= v_1 + a_1\Delta t\,, \end{aligned} \tag{2.7}$$

and the process can be *iterated* to give the equation of motion for any $t_i = t_0 + i\Delta t$:

$$\begin{aligned} a_i &= F(t_i, x_i, v_i)/m \\ t_{i+1} &= t_i + \Delta t \\ x_{i+1} &= x_i + v_i\Delta t \\ v_{i+1} &= v_i + a_i\Delta t\,. \end{aligned} \tag{2.8}$$

You might wonder whether you can trust your answer after many Δt's, and in fact, the solution obtained this way will "wander away" from the correct solution for larger values of t; but if you need improved accuracy, you must simply choose a smaller value of Δt. There is a trade-off, of course: a smaller Δt means more iterations before you get to the time you are interested in. Computer roundoff errors may also become important if too many iterations are required.

2.2.2 Second-Order Runge-Kutta

Euler's method is said to be *linear* in Δt because if you have an error in the position at some time t, then decreasing Δt by a factor of 2 will decrease the error by a factor of 2. More sophisticated methods for calculating the change in position and velocity after time Δt can improve the accuracy of the calculation and thus allow us to reduce the number of required steps. One described here is sometimes called the *second-order predictor-corrector* method, or the *second-order Runge-Kutta* method. The "second-order" in these names implies that the error made will decrease according to $(\Delta t)^2$, so that if you decrease Δt by a factor of 2, the error will decrease by a factor of $2^2 = 4$.

The method derives from Euler's method by recognizing that a better approximation to the new position is

$$x_1 = x_0 + \frac{1}{2}(v_0 + v_1)\Delta t\,. \tag{2.9}$$

This would be an improvement because the average value of the velocity accounts for the fact that the velocity is changing during the interval. From an integral point of view, the change in position is given by the area under the velocity versus time graph. Euler's method corresponds to a rectangular approximation to the integral, while Eq. 2.9 corresponds to the trapezoidal approximation.

A similar improvement can be written for the velocity:

$$v_1 = v_0 + \frac{1}{2}(a_0 + a_1)\Delta t. \tag{2.10}$$

Unfortunately, v_1 and a_1 are not so easy to find. We need a_1 to calculate v_1 in Eq. 2.10, but a_1 is determined by

$$a_1 = F(t_1, x_1, v_1)/m, \tag{2.11}$$

so we also need v_1 to be able to find a_1. The solution to this problem is to approximate x_1 and v_1 using Euler's method, use these values to calculate a_1, and then use Eqs. 2.9 and 2.10 to calculate improved values of v_1 and a_1. This is the origin of the *predictor-corrector* name for this method. To be specific:

$$\begin{aligned} x_1^{Euler} &= x_0 + v_0\Delta t \\ v_1^{Euler} &= v_0 + a_0\Delta t. \end{aligned} \tag{2.12}$$

These are the *predictor* values. These initial estimates for x_1 and v_1 are now used to calculate the new acceleration:

$$a_1^{Euler} = F(t_1, x_1^{Euler}, v_1^{Euler})/m, \tag{2.13}$$

and this acceleration is used to generate the more "correct" values of position and velocity using Eqs. 2.9 and 2.10:

$$\begin{aligned} x_1 &= x_0 + \tfrac{1}{2}(v_0 + v_1^{Euler})\Delta t \\ v_1 &= v_0 + \tfrac{1}{2}(a_0 + a_1^{Euler})\Delta t. \end{aligned} \tag{2.14}$$

While this is, of course, more complicated to work out for each time step, the advantage is that the error made in determining x_1 and v_1 is of order $(\Delta t)^2$. Thus, the total number of time steps should be much reduced.

2.2.3 Fourth-Order Runge-Kutta

The approach used in determining the second-order Runge-Kutta method may be extended to give progressively higher-order convergence to the correct result. For higher-order convergence methods, however, the work necessary to calculate each time step increases so that the reduction in the number of time steps may be offset by this added work. A reasonable compromise seems to be fourth-order Runge-Kutta. The derivation of the appropriate formulas may be found in most texts on numerical methods (see, for example, Press et al.[1] or Chapra and Canale[2]).

2.3 Using the Program

You should read the following section, The User Interface, and experiment with the predefined systems to give you some experience with the user interface. Then read the section on the basic user-defined procedures and experiment with defining your own system. Finally, consider the section on the animation procedures that show you how to take advantage of the full power of the Motion Generator. Exercises at the end of the chapter are noted according to the experience they require: (-) for exercises that use predefined systems, (+) for exercises that require programming a new force, and (*) for exercises that include animation. Once you are comfortable with adding animations, you will find that simple animations such as a vector showing the driving force will be helpful to include on many simulations.

The following terms are used in the Motion Generator in defining a system:

- *Dynamical Variables* are the positions, velocities, and accelerations. They are the time-dependent quantities used to represent the degrees of freedom of the system.

- *Force Parameters* are numbers used to define the forces. They can be changed by the user. If you define the force of gravity as $m*g$ rather than $m*(9.8\ N/kg)$ with g defined as a force parameter, then the user will be able to change the value of g during the simulation. The mass is usually included as a force parameter even though it is not strictly part of the force definition. A better name for force parameter might be "differential equation parameter."

- *User-Defined Functions* are functions of the dynamical variables that might be of interest to the user. For instance, the kinetic energy, potential energy, and angular momentum might be user-defined functions for a particular system. Note that these functions will usually also depend on the force parameters, as, for example, the angular momentum of a pendulum that depends on the length and mass as well as the angular velocity.

2.3.1 The User Interface

When a Motion Generator program is run, the program reads in a configuration file that provides initial values of the dynamical variables and the force parameters. This file also sets up the collection of windows to be used to illustrate the time evolution of the dynamical variables and functions. See Figure 2.1. All of these options may be changed and then saved to a new configuration file for later retrieval.

The function keys are used to control the simulation. **F2** starts the animation, **F3** pauses the motion and then steps through single time steps. **F4** reverses the motion. Note that when running *backward* the Motion Generator is not simply returning to previous values of the variables, but is integrating the equations of motion with a negative time step. This provides one means of investigating the convergence of the numerical procedure: you can run the simulation back to the start to see how close you come to your starting point. **F5** and **F6** control the speed of the animation by adjusting the time step between screen updates. **F10** takes you to a menu of options that can be used to modify the quantities controlling the simulation or to exit the program. Clicking with the mouse on the menu bar at any time will also take you into the menu. Clicking with the mouse anywhere during the simulation or pressing any key will make the animation pause. **F1** is always available to obtain context-sensitive help.

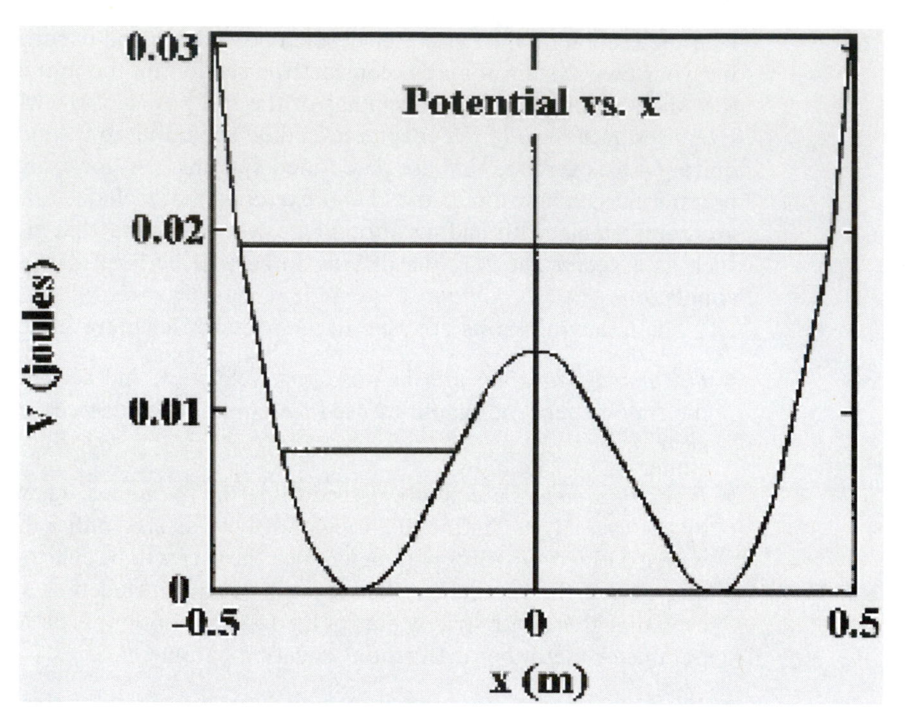

Figure 2.1: A GENMOT user screen (ElasPend).

Menu Options

- **Important Preliminaries:** You exit the program by selecting **Exit Program** under the **File** option in the menu. Help can be obtained by typing **F1** for context-sensitive help or by selecting **Help** from the menu. This option also allows you to view "user-defined" help that is specific to the particular Motion Generator program you are using.

- **File:** This menu option allows you to load and save files and exit the program. After determining a useful set of initial conditions, force parameters, and graph windows, you can save these as your own parameter file. You can also define your own help screen to accompany this parameter file. This can be a useful reminder of the purpose of a particular set of parameters or a source of hints to another user as to what to look for in running your simulation. You can also save the data you have generated to a file if you wish to do additional analysis of the motion.

- **Parameters:** This option allows you to change the force parameters, initial conditions, and numerical parameters. When changing the force parameters or the numerical parameters, you have the option of resetting the dynamical variables back to the initial conditions or continuing from the current values.

- **Graphs:** This option allows you to reset the dynamical variables to the initial conditions, to clear all graphs, or to change the color of the line drawn on the graph. Note that trajectories are not cleared automatically when parameters are changed, so that you can compare the motion of objects subject to different values of the force parameters or to different initial conditions.

- **Windows:** This option allows you to change the variables graphed in each window and to change the layout of the windows that are on screen during the simulation. You can add and delete windows, move them around, resize them, change their contents, and change the graph scales. The numerical window is the third line of the screen, giving numerical values of up to three variables. You can choose which variables to display numerically by selecting **Numerical Window**. You can also modify the window contents any time the simulation is paused by clicking with the mouse in the window. Similarly, clicking in the numerical window will allow you to change which quantities are displayed numerically.

- **Help:** This option provides a general help screen describing the Motion Generator, and a user-defined help screen that can be used to describe the particular configuration.

Before proceeding to the next section, you should run the sample program **GM1DGRAV**. This program describes the motion of an object in one dimension, subject to the force of gravity and an air resistance force that may be a linear and/or quadratic function of the velocity. This will give you some experience with the user interface of the program. Try changing initial conditions and values of parameters describing the force, then experiment with changing the graph variables and moving/resizing windows. You might want to work some of the exercises at the end of this chapter marked with a (-**GM1dGrav**), indicating that they can be completed with these sample programs.

You can also experiment with the other sample programs: Procedure **GMPend** generates the motion of a pendulum subject to a periodic driving force. Procedure **GMgener** is a system designed to represent many of standard force examples (i.e., spring, friction, gravity, periodic driving force). Procedure **GMrace** is a game designed to illustrate some of the animation capabilities of the Motion Generator. For some additional guidance, be sure to look at the user-defined help when you run these programs. You will also find exercises that use these sample programs.

2.3.2 The Basic User-Defined Procedures

The **GM1DGRAV** sample program will be used here to illustrate the Pascal procedures that describe a particular force. These procedures are found in **GM1DGRAV.PAS**, which contains additional comments.

The user-defined procedures defining a particular force are—

- **DefForceParms:** This procedure is called once when the program is run to generate the names, units, and descriptions of all of the quantities pertaining to the force, including dynamical variables, force parameters, and user-defined functions.

- **CalcAccel:** This procedure is called whenever the acceleration is needed. The procedure uses the current values of the dynamical variables and force parameters to calculate the acceleration.

- **CalcDynFun:** This function is called whenever one of the user-defined functions is needed. These user-defined functions can be any function of the dynamical variables and force parameters.

- **InitWindow:** This procedure is called whenever a window is opened. It can be used to draw initial graphics in a window. It is used in **GMrace**.

- **AnimateWindow:** This procedure is called each time a window is updated. It is used to generate animations. It is used in **GMpend**.

These procedures are discussed in more detail below.

- **DefForceParms**

  The variables **ForceTitle** and **ForceDesc** are string variables that hold descriptions of the force. **ForceTitle** is displayed at the top of the screen at all times, **ForceDesc** is displayed when the user modifies the parameters describing the force and should specifically describe the role of each parameter. Note that **ForceDef** is a data structure that holds all of the information about the force.

  Procedure **GenTimeVar** must be called just once to specify the name and unit of the time variable. Note that names are limited to 5 characters and units are limited to 10 characters in all the variable definitions. The names and units defined here will appear in the numerical window, on graphs, etc.

  Procedure **GenDynVar** must be called to describe the variables for each degree of freedom of the system (maximum of 3). For **1dGrav** there is only one degree of freedom. Note that only the unit of the position variable is given. The units for the derivatives are determined from this and the time unit.

  Procedure **GenForceParm** must be called to describe each parameter (maximum of 10) to be used in describing the force. Descriptions of these parameters are displayed when the user is asked to modify the parameters. These parameters are used in **CalcAccel** as described below. Note that a value of 9.8 N/kg for g could be used in calculating the acceleration in **CalcAccel**, but including it as a parameter allows the user to change its value during the simulation.

  Procedure **GenDynFun** is called to describe each of the user-defined functions that will be defined (maximum of 10). These functions may be used to calculate quantities such as the kinetic energy, angular momentum, or any function of the variables that may be of interest. These functions are calculated in **CalcDynFun** described below.

  The variable **ConfigFileName** holds the name of the default configuration file to be loaded when the program is run. If the file is not found when the program is run, the user is stepped through input screens to specify start-up values of the parameters and initial conditions.

```
PROCEDURE DefForceParms;FAR;
  BEGIN
    ForceDef.ForceTitle := '1-D Motion with Force of Gravity and Air
    Resistance';
    ForceDef.ForceDesc := 'F = -mg - c1 v - c2 v |v|';
   {+---------------------------------------------------------------}
   {|Specify name and units of "time" variable.
   {+---------------------------------------------------------------}
                 {Name }  {Units      }
    GenTimeVar('t',       's'           );
   {+---------------------------------------------------------------}
   {|Specify name and units of Dynamical Variables (Maximum of 3)
   {+---------------------------------------------------------------}
                 {Name }  {Vel    } {Acc   } {Units      }
    GenDynVar('x',        'vx',      'ax',     'm'           );
   {+---------------------------------------------------------------}
   {|Specify name and units of Force Parameters (Maximum of 10)
   {+---------------------------------------------------------------}
                   {Name }  {Units       } {Description          }
    GenForceParm('m',     'kg',          'Mass'               );
    GenForceParm('g',     'N/kg',        'Grav. Const'        );
    GenForceParm('c1',    'N/(m/s)',     'Linear Damping'     );
    GenForceParm('c2',    'N/(m/s)^2',   'Quadratic Damping'  );
   {+---------------------------------------------------------------}
   {|Specify name and units of User-defined Dynamical Functions
   {+---------------------------------------------------------------}
                   {Name }  {Units } {Description            }
    GenDynFun(1,'EK',     'Joules',     'Kinetic Energy'    );
    GenDynFun(2,'U',      'Joules',     'Potential Energy'  );
   {+---------------------------------------------------------------}
   {|Specify name of start-up parameter file
   {+---------------------------------------------------------------}
    ForceDef.ConfigFileName := 'GM1dGrav.PRM';
END;
```

- **CalcAccel**
  The main program calls this procedure whenever the current value of the acceleration is needed. When called, the current values of the dynamical variables and force parameters are available in the variables **x, vx, m, g, c1, c2,** which are used to calculate and return the value of the acceleration, **ax**. The other variables in the list, **y, vy, ay, z, vz, az, P5–P10,** are not used. The **y** and **z** dynamical variables are provided for problems with more than one degree of freedom, and **P5–P10** are unused force parameters. Note that the names of all these variables are arbitrary and should be defined to match the definitions given in **DefForceParms** (in the same order). Thus, of the 10 force-parameter variables **(P1–P10)**, only four are used in this simulation, and the names have been changed to **m, g, c1, and c2**.

```
PROCEDURE CalcAccel(VAR t,              {time variable}
                        x,vx,ax,        {x-dynamical variable}
                        y,vy,ay,        {Not Used}
                        z,vz,az,        {Not Used}
                        m,g,c1,c2,      {Force Parameters}
                        P5,P6,P7,
                        P8, P9,P10      {Unused Force Parameters}
                                  :Real );FAR;
VAR
  Force:Real;
BEGIN
  Force := -m*g - c1*vx - c2*vx*ABS(vx);
  ax := Force/m;
END;
```

- **CalcDynFun**
  The main program calls this procedure whenever a user-defined function is required for a graph or numerical display. When called, the variable **FunName** contains the name of the user-defined function as defined in **DefForceParms**, and the current values of the dynamical variables and force parameters are also passed to the procedure. As in the preceding procedure, the variable names have been chosen to match the definitions in **DefForceParms**.

```
FUNCTION CalcDynFun(FunName:String5; VAR
                    t,                       {time variable}
                    x,vx,ax,                 {x-dynamical variable}
                    y,vy,ay,                 {Not used}
                    z,vz,az,                 {Not Used}
                    m,g,c1,c2,               {Force Parameters}
                    P5,P6,P7,P8,P9,P10 {Not Used}
                              :Real ):Real; FAR;
BEGIN
  IF FunName='EK' THEN
    CalcDynFun := m*(vx*vx)/2
  ELSE IF FunName='U' THEN
    CalcDynFun := m*g*x;
END;
```

- **InitWindow,AnimateWindow**
  These two procedures can be used to generate animations. They are not used in **GM1DGRAV**. Refer to the next section for guidance on these procedures.

- **RunMotSim**
  The "main routine" simply calls a procedure in the **Gmutil** unit. The names of the user-defined procedures are passed as arguments to this routine to allow the unit to be compiled independently of the main routine.

```
{+--- This is the main routine.  RunMotSim is found in GMUtil ----+}
BEGIN
  RunMotSim(DefForceParms,CalcAccel,CalcDynFun,InitWindow,
  AnimateWindow);
END.
```

These three user-defined routines are all that are needed to produce a **GENMOT** program. Most of the information given in this section is included in comments in the actual **GM1DGRAV.PAS** file. Note that the two procedures **InitWindow** and **AnimateWindow** must also be included in this file even though they are not used.

2.3.3 The Animation Procedures

To illustrate the additional options available with the Motion Generator, consider the motion of a damped, driven pendulum of length L, with position and velocity described by the variables θ and ω. The equilibrium position of the pendulum is given by $\theta = 0$. The net torque acting on the pendulum may be written

$$\tau_{net} = -mg \sin\theta - c\omega + \tau_d \cos\theta \cos(\omega_d t), \tag{2.15}$$

where c is the magnitude of the dissipative term and the driving torque is applied with a frequency ω_d by a horizontal spring, attached to the pendulum partway down its length as shown in the Figure 2.2. The equation of motion is then determined from

$$\tau_{net} = I\alpha = (mL^2)\frac{d^2\theta}{dt^2}. \tag{2.16}$$

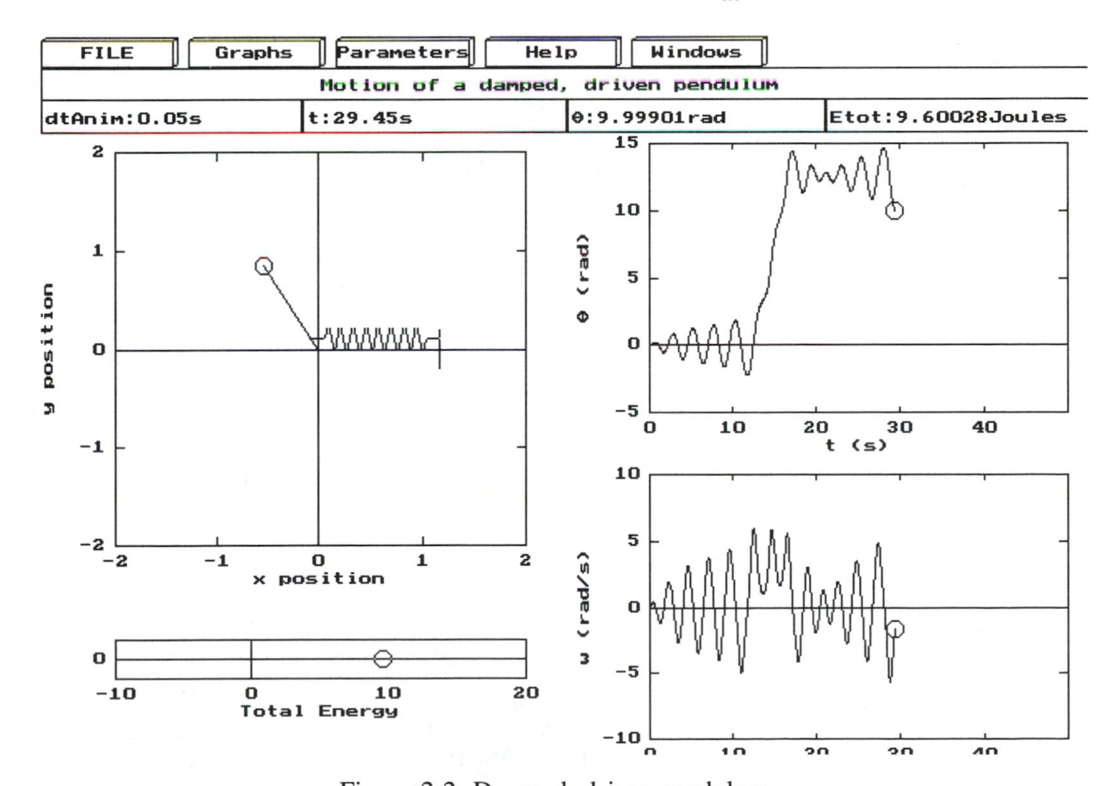

Figure 2.2: Damped, driven pendulum.

The routines defining this system are given below and may be found in **GMPEND.PAS**. Try running this sample program and then look at these routines to see how the animation is accomplished.

- **DefForceParms**

  The dynamical variable and its derivatives are θ, ω, and α. In IBM extended ASCII **CHR(233)** is θ and **CHR(224)** is α, so these characters could be used to represent the *names* of the variables. This is not done here to keep the program as simple as possible. The version of **GMPEND.PAS** included on disk *does* use the extended characters.

  Note that the user-defined functions, **xPos** and **yPos**, will be the position of the end of the pendulum in Cartesian coordinates. Inclusion of these functions will enable us to depict the actual motion of the pendulum bob.

```
PROCEDURE DefForceParms;FAR;
  BEGIN
    ForceDef.ForceTitle := 'Motion of a damped, driven pendulum';
    ForceDef.ForceDesc  := 'm  L^2  alpha = -mgL  sin(theta) - c1
    omega - T_d cos(w_d t)';

    GenTimeVar('t',      's'            );

    GenDynVar('theta','omega','alpha','rad'        );
    GenForceParm('m',     'kg',          'Mass'               );
    GenForceParm('g',     'N/kg',        'Grav Const'         );
    GenForceParm('c ',    'N/(rad/s)',   'Linear Damping'     );
    GenForceParm('L',     'm',           'Pendulum Length'    );
    GenForceParm('T_d',   'N-m',         'Drive Torque Mag'   );
    GenForceParm('w_d',   'rad/s',       'Drive Torque Freq'  );

    GenDynFun('EK',    'Joules',    'Kinetic Energy'    );
    GenDynFun('Etot',  'Joules',    'Total Energy'      );
    GenDynFun('xPos',  'm',         'x position'        );
    GenDynFun('yPos',  'm',         'y position'        );
    ForceDef.ConfigFileName := 'GMPend.PRM';
  END;
```

- **CalcAccel**

  This procedure generates the angular acceleration from the current values of **theta** and **omega**.

```
PROCEDURE CalcAccel(VAR
                      t,                        {time variable}
                      theta, omega, alpha,      {x-dynamical variable}
                      y, vy, ay,                {Not Used}
                      z, vz, az,                {Not Used}
                      m, g, c, L, T_d, w_d,     {Force Parameters}
                      P7, P8, P9, P10           {Not Used}
                                 :Real );FAR;
```

```
VAR
  Torque,MomentOfInertia:Real;
BEGIN
  MomentOfInertia := m*L*L;
  Torque :=   -m*g*L*SIN(theta) - c*omega +
  T_d*COS(theta)*COS(w_d*t);
  alpha := Torque/MomentOfInertia;
END;
```

- **CalcDynFun**

  As discussed above, **xPos** and **yPos** are used to plot the actual position of the pendulum bob in a window.

```
FUNCTION CalcDynFun(FunName:String5; VAR
                        t,                     {time variable}
                        theta,omega,alpha,     {x-dynamical variable}
                        y,vy,ay,               {Not Used}
                        z,vz,az,               {Not Used}
                        m,g,c,L,T_d,w_d,       {Force Parameters}
                        P7,P8,P9,P10           {Not Used}
                                 :Real ):Real; FAR;
BEGIN
  IF FunName='EK' THEN
    CalcDynFun := m*Pwr(omega*L,2)/2
  ELSE IF FunName='Etot' THEN
    CalcDynFun := -m*g*L*COS(theta) + m*Pwr(omega*L,2)/2
  ELSE IF FunName='xPos' THEN
    CalcDynFun := L*SIN(theta)
  ELSE IF FunName='yPos' THEN
    CalcDynFun := -L*COS(theta);
END;
```

- **AnimateWindow**

  In the plot of **yPos** versus **xPos** for this system, it would be helpful in visualizing the system to draw the arm of the pendulum. This can easily be accomplished in the procedure **AnimateWindow** which is called every time any window is updated. The variables **xAxisVar** and **yAxisVar** contain the names (as defined by **DefForceParms**) of the variables being plotted. Thus, if this window corresponds to **yPos** versus **xPos**, the procedure draws a line from the pendulum bob to the origin.

```
PROCEDURE AnimateWindow(yAxisVar,xAxisVar:String5;
Erase:Boolean; VAR
                        t,              {time variable}
                        theta,omega,    {x-dynamical variable}
                        y,vy,ay,        {Not Used}
                        z,vz,az,        {Not Used}
                        m,g,c,L,T_d,
                        w_d,            {Force Parameters}
                        P7,P8,P9,P10    {Not Used}
                              :Real ); FAR;
```

```
VAR
  xPos,yPos:Real;
  x1,x2,y1:Real;
BEGIN
  IF (yAxisVar='yPos') AND (xAxisVar='xPos') THEN
    BEGIN
      xPos := L*SIN(theta);
      yPos := -L*COS(theta);
      PlotLine(0,0,xPos,yPos);
    END;
END;
```

Note that the user-defined functions for **xPos** and **yPos** could be used in this procedure by calling **CalcDynFun('xPos',....)** and **CalcDynFun('yPos',...)** but the definitions are sufficiently simple that it is easier to just reproduce them here. **PlotLine** is a CUPS procedure that draws a line in a window using the current scale. Three other useful procedures are **DrawSpring(x1,y1,x2,y2,width)**, which draws a spring of width determined by **width** between the two pairs of points; **DrawVector(x1,y1,Vx,Vy)**, which draws a vector with components **Vx,Vy** starting at the point **(x,y)**, and **DrawCircle(x1, y1)**, which puts a circle on the screen.

The procedure **AnimateWindow** is actually called twice each time a window is updated: the first time, the procedure erases the old image on the screen; the second time, the image at the new position is drawn. The Pascal **WriteMode** is set to **XORput** when this routine is called, which allows **AnimateWindow** automatically to erase and redraw without any additional programming. The variable **Erase** is included, however, so that your procedure can determine whether it is being called to erase an old image or create a new one.

It also might be useful to draw explicitly the spring that is generating the external torque. The additional lines of code shown below will generate the required animation. **x1** and **y1** are the positions along the pendulum at which the spring is attached. **x2** is the position of the other end of the spring that is oscillating at a frequency of **w_d** to provide the torque. The line plotted at **x2** helps the user visualize this end of the spring.

```
PROCEDURE AnimateWindow(...); FAR;
VAR
  xPos,yPos:Real;
  x1,x2,y1:Real;
BEGIN
  IF (yAxisVar='yPos') AND (xAxisVar='xPos') THEN
    BEGIN
      xPos := L*SIN(theta);
      yPos := -L*COS(theta);
      PlotLine(0,0,xPos,yPos);
      IF T_d<>0 THEN
        BEGIN
```

```
        x1 := xPos/8;
        y1 := ypos/8;
        x2 := 1.2 + 0.2*COS(w_d*t);
        DrawSpring(x1,y1,x2,y1,0.1);
        PlotLine(x2,0.2,x2,-0.2);
      END;
   END;
END;
```

InitWindow
This procedure works very much like **AnimateWindow**. It is called when a window is initially opened and whenever all graphs are cleared. It can be used to draw static images that you want present throughout the simulation. Examples might be electric or magnetic field lines.

2.4 Exercises

Exercises are marked according to the experience they require: (-) for exercises that use predefined systems; (+) for exercises that require programming a new force; and (*) for exercises that include animation.

2.1 (-**GM1dGrav**)(Comparison of Numerical Methods). Consider the motion of a 1.0 kg object in one-dimension (1-D) subject to a resistive force proportional to the velocity of the object:

$$F = -c_1 v, \tag{2.17}$$

where c_1 has the value 0.5 N/(m/s). Note that there is no force of gravity for this problem, so you should set $g = 0$. The object starts at the origin at $t = 0.0$ s and is given an initial velocity of 300 m/s.

(a) Solve equation Eq. 2.17 for $x(t)$ and show that the position of the object after 3.0 s is 466.122 m.

(b) Use the Motion Generator program **GM1dGrav** with the Euler method to describe this motion. Use a value of dt equal to 0.1 s and determine the numerical value of the position at $t = 3$ s. Determine the percent error based on the value from part a.

(c) Repeat part b with $dt = 0.01$ s, 0.001 s and 0.0001 s. Make a table showing your values, including the percent error. Note that you should come very close to the correct value from part a by the time you get to $dt = 0.0001$ s.

(d) Now change the numerical method in **GM1dGrav** to **RK2** and repeat parts b and c.

(e) Determine the *order of the convergence* in each case. That is, determine the exponent p such that the percent error is given by

$$err = A\,(dt)^p. \tag{2.18}$$

You should be able to determine p simply by looking at the data. Determine the value of A in each case as well.

(f) Now change the numerical method in **GM1dGrav** to **RK4** and repeat part b. How large a value of dt can you use and still obtain 6 digits of accuracy? Note that values of **dt** larger than **dtAnim** are ignored since the program must evaluate the positions and velocities every **dtAnim** seconds in order to update the screen.

Note: For *every* numerical problem you must *always* check convergence by investigating the effect of changing dt. A value of dt that is sufficient for one set of initial conditions may not suffice for another set. Moral: Be careful when doing numerical work.

When using the Motion Generator in subsequent problems you should state the numerical procedure used, the value of dt, and the accuracy of reported values (careful use of significant digits should suffice).

2.2 (-**GM1dGrav**)(Exponential Decay). Consider the motion of a 0.25 kg object in 1-D subject to a resistive force proportional to the speed of the object:

$$F = -c_1 v, \tag{2.19}$$

where c_1 has the value 0.2 N/(m/s). Note that there is no force of gravity present for this problem.

(a) Use the Motion Generator program **GM1dGrav** to determine the distance traveled by the object in 10 s if it is given an initial velocity of 10 m/s.

(b) Determine the answer to part a analytically and compare your answers.

(c) If the object is initially traveling at 10 m/s, determine the time required for the object to reach a speed of 5 m/s and the distance traveled during this interval. Note carefully the number of significant digits in your values. Now determine the time interval required for the object to go from 5 m/s to 2.5 m/s and from 2.5 m/s to 1.25 m/s. What is the distance traveled *during* each of these intervals? Make a table showing $x(t)$ and $v(t)$ as well as Δx and Δv for each time interval. Discuss the relationships between successive values of Δx and Δv.

(d) From your table from part c, estimate the time required for the object to reach a speed of 0.01 m/s, and the total distance traveled at this point.

Note: Be sure to specify the numerical algorithm you use with **1dGrav** and the value of dt, including a justification for your choice of the size of dt.

2.3 (-**GM1dGrav**). Consider the motion of a 0.25 kg object in 1-D subject to the force of gravity and a resistive force proportional to the speed of the object:

$$F = mg - c_1 v. \tag{2.20}$$

(a) Use the Motion Generator program **GM1dGrav** to determine the value of c_1 accurate to two decimal places that generates a terminal velocity of 2.0 m/s. Then determine the value analytically and compare your answers. Use the analytic value for the remainder of the problem.

(b) Use **GM1dGrav** to determine the time required for this object to reach the ground if it is dropped from a height of 20 m. What is the speed of the object when it reaches the ground?

(c) Suppose the object is thrown downward with a speed of 15 m/s from a height of 20 m. Determine the time required for the object to reach the ground in this case, and the speed of the object.

(d) Which of the results from parts b and c can be determined analytically? Determine these values and compare them to your numerical results.

2.4 (-**GM1dGrav**). A 0.25 kg object is thrown vertically upward with an initial velocity of 25 m/s. The gravitational constant has a value of $g = 9.8$ N/kg.

(a) Calculate the maximum height, the time required to reach this height, and the time required for the object to return to the starting point, ignoring air resistance.

(b) Suppose a drag force is also present given by

$$F_{air} = -c_2 v|v|, \tag{2.21}$$

where $c_2 = 0.05$ N/(m/s)^2. Which quantities calculated in part a will be different? Will they be smaller or larger? Will the time for the object to reach maximum height be equal to the time for the object to return? Try to answer these questions using your intuition, and write short explanations justifying your responses. *Then* use the Motion Generator program **GM1dGrav** to test your answers and record the numerical results.

2.5 (-**GMPend**) (Sensitivity of Non-linear Systems to Initial Conditions). The system investigated here is described in section 2.3.3 of the text and is available in the Motion Generator program **GMPend**. Consider a mass of 1.0 kg with a damping coefficient $c = 0.1$ N-m/(rad/s), and external torque of frequency 2.8 rad/s with amplitude of 2.0 N-m. The initial value of θ is 0.03 rad and the initial angular velocity is zero.

(a) Set up graphs to show θ versus t and ω versus t over the interval $0 \leq t \leq 50$ s. Run the simulation to determine the approximate value of dt necessary to generate $\theta(t)$ to an accuracy of 4 decimal places at 24 s. How accurate is your value for θ at 50 s with this value of dt? Record the values of θ at 10 s and at 24 s. Is this motion periodic?

(b) Change the initial value of θ to 0.031 rad and rerun the simulation (you should change the color of your graph to help you look for deviations of the motion from the previous result). By how many radians is the value of θ changed (compared to results with $\theta_0 = 0.03$ rad) at 10 s and at 24 s? Repeat this with initial values of θ equal to 0.032 rad and 0.033 rad. How much variation in the curves for $\theta(t)$ and $\omega(t)$ do you see for these different initial conditions?

(c) Change the magnitude of the external torque to 3 N-m, and clear the graphs. Repeat the analysis above, starting with $\theta_0 = 0.03$ rad. Is the motion periodic? Consider the value of θ at 10 s and 24 s with $\theta_0 = 0.031$ rad and then $\theta_0 = 0.0301$ rad. You should recheck your value of dt to make certain that the effects you see are not the result of numerical imprecision. Can you find a small enough shift in the value of θ_0 that gives you essentially the same motion for 50 s?

2.6 (-**GMRace**) (A Day at the Races). The arrow keys can be used in this simulation to provide a force on the "racecar." Try to complete a circuit around the track in as short a time as possible, staying within the track. When you get a satisfactory run, print a copy of the screen.

Obvious Hints: If the acceleration is perpendicular to the velocity, the direction changes, but the speed does not. If the acceleration is "tilted" forward, the speed increases; if backward, the speed decreases.

2.7 (-**GMgener**). Consider the following force given as a function of t:

$$F_x(t) = F_0 - a_1 t, \tag{2.22}$$

where $F_0 = 20$ N and $a_1 = 10$ N/s. This force is applied to a 1.0 kg object that is initially at rest at the origin ($x = 0$). This system may be simulated using the Motion Generator program **GMgener**. Note that you can use mg in **GMgener** to generate F_0.

(a) Determine the analytic expression for $x(t)$ for this system.

(b) Use **GMgener** to generate a graph of $x(t)$ for the range $0 \le t \le 6$ s. Determine the maximum value of x and the time at which this value is reached. Compare the numerical results with your analytical results.

(c) Use **GMgener** to find an initial velocity that will bring the object back to the origin in 10 s. An accuracy of two digits in the initial velocity is sufficient for this part. Make a rough sketch of the position versus time graph for this motion.

(d) Use **GMgener** to find an initial velocity that will bring the object back to the origin after 4 s. An accuracy of 3 digits is needed for this part. Make a rough sketch of the position versus time graph for this motion.

2.8 (-**GMgener**)(Oscillatory Motion). Consider the three different potentials, $U_i(x)$, $i = 1, 2, 3$, given by

$$\begin{aligned} U_1(x) &= A_1|x| \\ U_2(x) &= A_2x^2 \\ U_3(x) &= A_3x^4. \end{aligned} \tag{2.23}$$

All three of these potentials will generate periodic motion. The goal of this problem is to determine the dependence of the period on amplitude, mass, and the parameters A_i.

(a) Draw a rough sketch of the potential for each case. In your sketch, indicate clearly how U_2 and U_3 are different. Determine the force corresponding to each potential (be careful about direction for F_1).

(b) The Motion Generator program **GMgener** can be used to describe each of these potentials. For each potential, investigate the relationship between the period, T, and the amplitude of the motion x_m by picking a value for the mass and for the A_i (record the values you use) and then experimenting with various initial positions (use $v_0 = 0$). Try amplitudes whose ratios are small integers. For each potential you should find

$$T = C(x_m)^p, \tag{2.24}$$

where C and p are constants. Estimate the value of p for each case. Note that you can use a plot of $\log_{10}(T)$ versus $\log_{10}(x_m)$ to do this if you cannot determine p by inspecting the data.

(c) For a fixed amplitude and mass investigate the dependence of the period on the A_i by running **GMgener** with different values of A_i. Do this for each potential.

(d) For a fixed amplitude and A_i, investigate the dependence of the period on the mass, m, by running **GMgener** with different values of m. Do this for each potential.

(e) Use the results from parts b–d to determine the function giving the period in terms of all of the parameters in the problem. That is

$$T = f(m, A_i, x_m) \tag{2.25}$$

for each case.

2.9 (+) (Washboard Potential). Consider the following force acting on a 1.0 kg object given as a function of position:

$$F_x(x) = F_0 \cos(2\pi x/L) - cv, \tag{2.26}$$

with $F_0 = 10$ N, $L = 2$ m, and $c = 0$ initially.

(a) Draw a graph of the potential energy corresponding to the first term of this force.

(b) Modify a Motion Generator program to describe this force. Include m, F_0, L, and c as parameters and set up user-defined functions to give the potential energy and the total energy. Run the program and display the numerical value of the total energy to check your work. Submit a copy of your program.

(c) Suppose that the object is initially at rest at $x = 0.0$ m. Use your Motion Generator program to determine the subsequent motion. Make a rough sketch of the position versus time graph and the velocity versus time graph and describe the qualitative features. What is the period of the motion?

(d) Repeat part c with an initial velocity of 2.5 m/s (check your value of *dt!*). Discuss, qualitatively, the differences between the curves from parts c and d. What will happen for larger values of the velocity? Explain your answer using your sketch of the potential.

(e) Add a graph to the output window to show the potential energy of the object as a function of position (note that you will need to move the other graphs around to make room on the screen). Rerun the simulation from part d and print a copy of the screen.

(f) Add a graph to the output window of the Motion Generator program to show the velocity of the object versus the position (i.e., velocity on the vertical axis and position on the horizontal axis). This kind of graph is called a phase plot. Rerun the simulation from parts c and d and describe the motion of the point on this phase plot. Print a copy of the screen.

(g) Add a frictional term to this force (use $c = 0.5$ N/[m/s]) and rerun the simulation from parts c and d. Experiment with larger values of the initial velocity, and determine the range of initial velocities that will allow the object to move to the next "bump" and stay there. Run the simulation with one of these values and print a copy of the screen. Make certain that you understand the phase plot.

2.10 (+) (Impulsive Forces). A 1.0 kg object is constrained to move in 1-D and is initially at rest at the origin. The following time-dependent force acts on this object:

$$F(t) = \frac{A}{\sigma} exp \left[\frac{-(t - t_0)^2}{\sigma^2} \right], \tag{2.27}$$

where t_0, A and σ are fixed parameters, initially set at $t_0 = 5$ s, $A = 10$ N $-$ s, and $\sigma = 1$ s.

(a) Draw a sketch of $F(t)$ over the time interval $0 \leq t \leq 10$ s. Draw additional curves for $\sigma = 0.5$ s and $\sigma = 0.3$ s as well.

(b) Modify a Motion Generator program to describe this force, allowing m, t_0, A, and σ to be adjustable parameters. Change the user-defined functions to include the kinetic energy and the value of the force, which can be generated as m times the acceleration. Note that there is no potential energy associated with this force. Submit a copy of your program.

(c) Use your program to generate a graph of velocity versus time for $0 \leq t \leq 10$ s for values of σ equal to 1.0 s, 0.5 s and 0.3 s. Print a copy of the screen and describe *in detail* how and why the three curves are similar and different. Hint: The final velocities should be the same in all three cases. Check and report on numerical methods and parameters.

(d) Add two more graphs to the motion simulator to display the force versus time and the force versus distance, after removing the position versus time graph. Rerun all three examples from part c, and describe the relationship among the three graphs. Recall that the change in momentum is given by the integral (the area under the curve) of the graph of $F(t)$, while the change in kinetic energy is given by the integral of the graph of $F(x)$.

2.11 (*) (Projectile Motion). Consider the motion of an object in two dimensions, subject to the force of gravity and a linear air resistance force:

$$\begin{aligned} F_x &= -c_1 v_x \\ F_y &= -mg - c_1 v_y . \end{aligned} \tag{2.28}$$

(a) Modify a Motion Generator program to describe this system. Define dynamical functions to generate the kinetic energy, the potential energy, and the total energy. Run the program with $c_1 = 0$ for an object of mass 1.0 kg given an initial velocity of $v_{x0} = 20$ m/s and $v_{y0} = 20$ m/s and verify that energy is conserved. Determine, analytically, the range of the projectile and verify that you obtain the same result with your Motion Generator program.

(b) Use a value for c_1 that produces a terminal velocity of 100 m/s. Note that you can determine this value analytically, but you should check that it works with your Motion Generator program. Use your program to determine the range and time aloft when the object is launched with the same initial velocity given in part a. Show that this result agrees with the exact equations,

$$\begin{aligned} x(t) &= \frac{v_{x0}}{\gamma}(1 - e^{-\gamma t}) \\ y(t) &= \left(\frac{v_{y0}}{\gamma} + \frac{g}{\gamma^2}\right)(1 - e^{-\gamma t}) - \frac{g}{\gamma} t, \\ \gamma &= c_1/m \end{aligned} \tag{2.29}$$

by plugging in the time aloft from your Motion Generator program. How well does the approximate range equation

$$x_r \approx \frac{2 v_{x0} v_{y0}}{g} - \frac{8}{3} \frac{v_{x0} v_{y0}^2 \gamma}{g^2} \tag{2.30}$$

work for this problem?

(c) Modify your Motion Generator program to draw the acceleration vector attached to the object when a graph of y versus x is drawn. Refer to section 2.3.3 for guidance. How does the acceleration vector change when air resistance is present?

(d) Modify your Motion Generator program to draw the tangential and normal components of the acceleration vector attached to the object when a graph of y versus x is drawn. What does the tangential component of the acceleration tell you about the motion? Submit a copy of your program (with the .PRM file) on disk.

2.12 (*) (Two-Dimensional Harmonic Oscillator)

(a) Modify a Motion Generator program to describe the motion of an object in two dimensions, subject to a linear restoring force given by

$$F_x = -kx$$
$$F_y = -ky\,. \tag{2.31}$$

Use cgs units with the mass of the object equal to 100 g and the spring constant $k = 500$ dynes/cm. Define dynamical functions to generate the kinetic energy, the potential energy, and the total energy (in ergs). Note that you should make certain that the program displays the correct units throughout. Run the program with initial conditions given by $x_0 = 5$ cm, $y_0 = 5$ cm, $v_{x0} = v_{y0} = 0$ and determine the frequency in cycles per second. Check that total energy is conserved.

(b) Add springs to the graph of y versus x to show explicitly the harmonic force. You should draw four springs attached to the mass with the four other ends connected to the points (-10 cm, 0), (10 cm, 0), (0, -10 cm), (0, 10 cm). Submit a copy of your program (with the .PRM file) on disk.

(c) Determine the value of v_{y0} required for the object to travel in a circle of radius 5 cm if the remaining initial conditions are given by $x_0 = 5$ cm, $y_0 = 0$ cm, $v_{x0} = 0$. Do this analytically by remembering that the force required to make an object go in a circle of radius R is given by

$$F_{net} = mv^2/R\,. \tag{2.32}$$

What is the period of the motion? What happens if the initial y velocity is less than or greater than the value for circular motion? Describe this motion. Is it still periodic?

(d) Modify your Motion Generator program to describe a force given by

$$F_x = -kx(x^2 + y^2)$$
$$F_y = -ky(x^2 + y^2)\,, \tag{2.33}$$

with $k = 20$ dynes/cm^3. You should eliminate the spring animation since this force is no longer harmonic. Define dynamical functions to generate the energy. Note that in cylindrical coordinates, this force is given by

$$\vec{F} = -kr^3\hat{e}_r\,, \tag{2.34}$$

so the potential energy is given by

$$U = \frac{1}{4}kr^4 = \frac{1}{4}k(x^2 + y^2)^2\,. \tag{2.35}$$

Repeat part c for this force (check that total energy is conserved). Consider initial velocities that deviate by ≈ 20.

Note: The remaining problems all require Lagrangian mechanics.

2.13 (*) (Elastic Pendulum). Consider a mass, m, attached to a spring with force constant k and equilibrium length L_0. The mass is free to move in two dimensions (up-and-down and side-to-side).

(a) Set up the Lagrangian for this system with θ and μ as the generalized coordinates, where θ is the displacement of the pendulum from equilibrium and μ is the stretch of the spring away from its equilibrium position, $L_0 + mg/k$. Determine the differential equations for θ and μ from Lagrange's equations.

(b) Modify a Motion Generator program to describe this system, with m, k, and L_0 as modifiable force parameters. Set up user-defined functions to generate the kinetic energy, total energy, and the x and y positions of the mass. Use the **AnimateWindow** procedure to draw the spring in the y versus x window. Check that your program gives reasonable results and conserves energy. Submit a copy of your program (with the .PRM file) on a disk.

(c) For small-amplitude motion, the differential equations can be *decoupled* to give the two expected modes (simple harmonic motion [SHM] of the spring, and SHM of the pendulum). Show that this is the case from your Lagranian and determine the expected frequencies. Use your imulation with $k = 20$ N/m, $L_0 = 1.0$ m, and $m = 1.0$ kg to show that for small initial θ (try $\theta_0 = 0.1$ rad) the motion is essentially pendulum-like (that is: not much oscillation of the spring). Determine the period of the motion.

(d) Change the value of L_0 above to 1.47 m and rerun the simulation with small initial θ. How is the motion different? Why are the two modes no longer decoupled? Hint: Determine the frequencies of pure pendulum and pure spring motion. How are these related?

2.14 (*) (Linear Pendulum). A pendulum of mass m_2 and length L_0 is attached to a block of mass m_1 that is free to slide without friction along a horizontal rail. The block is tied to a rigid support by a spring with spring constant K.

(a) Determine the Lagrangian of the system using the following dynamical variables: x, the displacement of the block; and θ, the angle that the pendulum makes with the vertical.

(b) Use Lagrange's equations to determine the differential equations of motion of the system. Solve these equations (algebraically) for $\ddot{x}$ and $\ddot{\theta}$.

(c) Modify a Motion Generator program to describe this system using the differential equations obtained in part b. Include dynamical functions in your program to generate the x and y positions of the pendulum. Modify the **animateWindow** procedure to draw the system in the y versus x window. Run the program for $\theta_0 = \pi/2$ rad using $m_1 = 2$ kg, $m_2 = 0.5$ kg, $L_0 = 0.5$ m, and $K = 20$ N/m. Check that the motion makes sense and that energy is conserved. Submit a copy of your program (with the .PRM file) on disk.

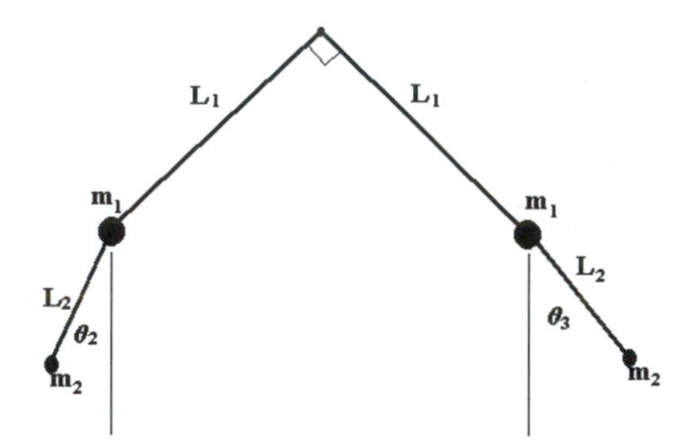

Figure 2.3: Chaotic Pendulum.

(d) Consider small oscillations of the system. Show that if x and θ are small, the differential equations may be written in the form of a system of coupled oscillators (i.e., find the dynamical matrix describing the motion). Insert the specific numerical values for the parameters in the dynamical matrix and solve for the normal mode frequencies. Determine the relationship between x_0 and θ_0 for each normal mode and check that you obtain simply harmonic motion at the correct frequencies in your Motion Generator program with these initial conditions. Record the initial values of x and θ you use in each case.

2.15 (*) (Chaotic Pendulum) Note: This problem is more appropriate as a class project. Two identical objects of mass m_1 are connected to the ends of light rods of length L_1 whose other ends are connected together at a fixed angle of 90°. See Figure 2.3. This object is suspended and is free to pivot about its 90° corner. A light rod of length L_2 is suspended from each of the m_1 objects and identical objects of mass m_2 are connected to the free ends of these rods. The entire system is constrained to motion in a single plane. The variable θ_1 is used to represent the motion of the m_1 objects away from their equilibrium position. The variables θ_2 and θ_3 measure the rotation of the L_2 rods from the vertical axis. Note that all three angles are zero at equilibrium.

(a) Determine the Lagrangian of the system.

(b) Use Lagrange's equations to determine the differential equations of motion of the system. Solve these equations (algebraically) for $\ddot{\theta}_1$, $\ddot{\theta}_2$, and $\ddot{\theta}_3$.

(c) Modify a Motion Generator program to describe this system using the differential equations obtained in part b. Include dynamical functions in your program to generate the x and y positions of one of the m_1 objects. Modify the **animateWindow** procedure to draw the system in the y versus x window. Run the program using $m_1 = 2$ kg, $m_2 = 0.5$ kg, $L_1 = 0.9$ m, and $L_2 = 0.3$ m. Check that the motion makes sense and that energy is conserved. Submit a copy of your program (with the .PRM file) on disk.

(d) Consider small oscillations of the system. Show that if all angles are small, the differential equations may be written in the form of a system of coupled oscillators (i.e., find the dynamical matrix describing the motion). Insert the specific numerical values for the parameters in the dynamical matrix and solve for the normal mode frequencies. Determine the relationship between the initial angular displacements for each normal mode and check that you obtain simply harmonic motion at the correct frequencies in your Motion Generator program with these initial conditions. Record the initial values of the angles you use in each case.

References

1. Press, W. H., Flannery, B. P., Teukolsky, S. A., Vetterling, W. T. *Numerical Recipes in Pascal*. Cambridge: Cambridge University Press, 1989.

2. Chapra, S. C., Canale, R. P. *Numerical Methods for Engineers*. New York: McGraw-Hill Book Company, 1988.

3

Anharmonic Oscillators

Bruce Hawkins

Next, when I cast mine eyes and see
That brave vibration each way free;
Oh how that glittering taketh me!

Robert Herrick, *Upon Julia's Clothes*

3.1 Introduction

A quick introduction to the features of the computer program is in the ANHARM walk-through in Appendix A.

Oscillators are ubiquitous in all realms of nature, from the atomic to the astronomical, from the biological to the technological. Your legs and arms behave as oscillators when you walk; your heart is an oscillator; an automobile has oscillating parts (obvious when you drive over a rough road); a clock is regulated by an oscillator, whether it be a pendulum or a quartz crystal; atoms oscillate; and stars pulsate. The simple harmonic oscillator described in first-year physics is only an approximation to any of those just mentioned; it reveals fundamental aspects of the behavior of all of them, yet leaves unmentioned other important aspects.

Of all of these oscillators, the clock is the closest to exhibiting simple harmonic behavior, and the clock designer's task is to minimize the effect of anharmonicity. Yet the clock requires complexity, for simple harmonic oscillators, being frictionless, do not exist in nature. The clock requires a constant supply of energy to keep it running. This energy supply introduces the seeds of chaos, which chaos the clock designer must hold at bay. Fortunately this is not hard in the case of the clock. The Designer of the animal heart had a more difficult task, for the seeds of chaos are closer. Many now think they are essential to its proper operation, yet uncontrolled they lead to the death of the owner.

3.2 Theoretical Concepts

The basic concepts of oscillatory motion are treated in many textbooks.[1–12] These concepts include period, damping and decay, driving forces, resonance, and the interactions of all these. Even though some textbooks discuss some of the effects of nonlinearity (anharmonicity), these effects are so rich as to make full coverage impossible in a course that must discuss other topics.

The computer simulation described here is designed to help you understand the material covered in your text, as well as introduce you to a number of concepts not covered there. The simulation also has a tutorial section that is intended to introduce you to a number of concepts.

3.2.1 Phase Space

This is a concept that often appears mysterious when first encountered. There is no mystery to the fact that we can plot a graph of v against x if we wish to do so. This is called a phase space plot. The origin of the term *phase space* is probably in orbital mechanics, which deals, among other things, with the phase of the orbit. But why should we be interested in such a plot?

One reason is to show patterns that can be seen in no other way. Surprising regularities appear for seemingly irregular motions. Another is that there exist powerful theorems about motion in phase space which allow us to draw conclusions without actually following the motion of a system. That is the opposite of what we are doing when we use a simulation like this which follows the motion. The two approaches complement each other. The tutorial displays a number of phase plots designed to call your attention to what can be learned from them.

3.2.2 Poincaré Diagram

Nearly one hundred years ago Henri Poincaré proposed that insight can be gained by observing the position in phase space at regular time intervals, spaced according to some time relevant to the particular system. When dealing with a driven oscillator, a time interval likely to be useful is one period of the applied sinusoidal driving force. These positions are then all plotted. The resulting plot is called a Poincaré diagram. If a trajectory in phase space is periodic, it is apparent in such a diagram even if the period is so long that it is not otherwise evident.

Non-periodic trajectories which appear to fill all of the phase diagram within certain limits may show intricate structure in the Poincaré diagram. These are called strange attractors because the trajectory is attracted to certain portions of phase space, yet is strange compared to the periodic orbits with which we are familiar. The nature of such a structure is better seen than described.

Even more information about the structure of a strange attractor can be obtained by linking into a movie a number of Poincaré diagrams made at different phases of the driving force. That is to say, the first Poincaré diagram is a plot of (v, x) pairs taken at $t = 0,\ T,\ 2T, \ldots,$ the next at $t = T/n,\ T + T/n,\ 2T + T/n, \ldots,$ the next at $t = 2T/n,\ T + 2T/n,\ 2T + 2T/n, \ldots$ and so on.

3.2.3 Return Diagram

Another way of revealing a strange attractor is to plot successive pairs of the values of some variable, for example, v_n, v_{n-1} pairs, where v_n and v_{n-1} are the velocities at the nth and $n - 1$st time steps. This can be done with experimental data for any variable even if the relevant time interval is not known, and even in some cases if the data are not collected at regular time intervals. A case in point is the times at which water drops fall from a dripping faucet. Here the time interval, while irregular, is meaningful, and it is the successive time intervals themselves that are plotted.

3.2.4 Hooke's "Law"

Hooke's "law" is a reasonable approximation to the behavior of real springs, but no spring follows it perfectly. If nothing else, all springs break if you stretch them too far. A spring is said to be *hard* if its spring constant increases as it is stretched, and *soft* if the constant decreases. A pendulum behaves like a soft spring. Most springs get hard if you try to compress them very far, or they may buckle, thus becoming soft.

3.2.5 Attractors and Basins of Attraction

An undriven oscillator with friction will gradually come to rest at its equilibrium position. It is said to have a *point attractor* at the equilibrium position. The trajectory in phase space will spiral in to the equilibrium position at x_0, 0, as shown in Figure 3.1. A clock pendulum has friction and is driven by a mechanism in such a way that it oscillates at a constant amplitude as long as someone remembers to wind the clock. Nowadays, we describe this system as having a *periodic attractor.* If you poke the pendulum to either increase or decrease its amplitude, it soon returns to its preferred amplitude. It seems to be "attracted" to one particular motion, whose track in phase space is a closed curve. This closed curve is called a *limit cycle*. Figure 3.2 shows the behavior of a driven pendulum started from rest. It does not simply spiral toward the limit cycle but wanders back and forth across it before settling down. If you decrease the pendulum amplitude too much, however, many clocks will stop. Now we have two attractors: a point and a limit cycle. Depending on how you start the clock, it is attracted to either one or the other. If you patiently start the clock pendulum over and over again, plotting all the starting points in phase space and coloring them either red or blue depending on which attractor the motion goes to, you will find that the red and blue points divide phase space into two parts. These are called the *basins of attraction* of the two attractors.

In this simple case, the basin of attraction to the rest point probably really is shaped more-or-less like a wash basin. The other is all the rest of the space. Systems can have more than two basins, and the shapes of the basins need not be simple; they can be entwined with each other. Understanding these concepts is best achieved by searching for some basin boundaries (Exercise 6).

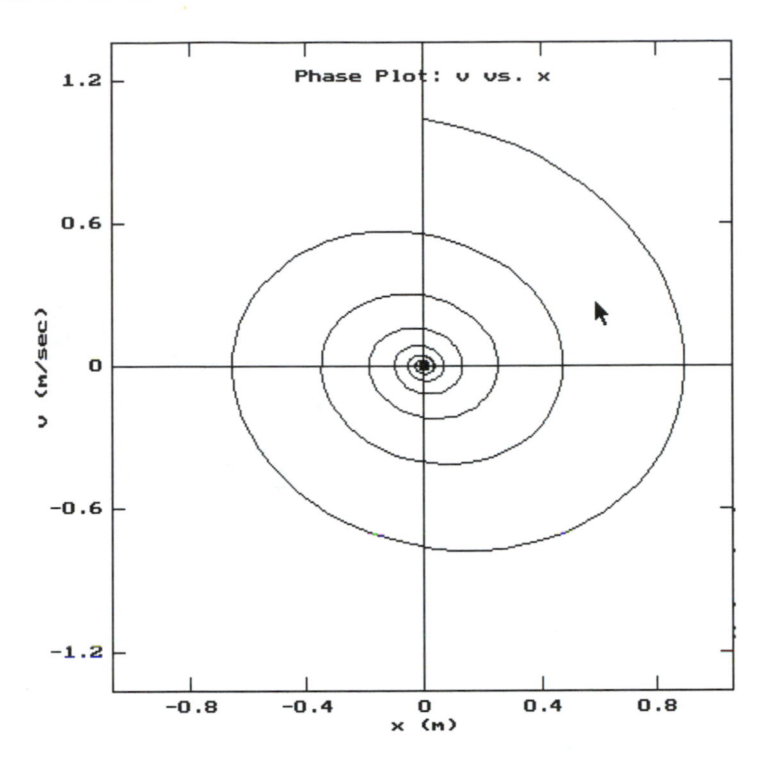

Figure 3.1: Effect of friction in phase space.

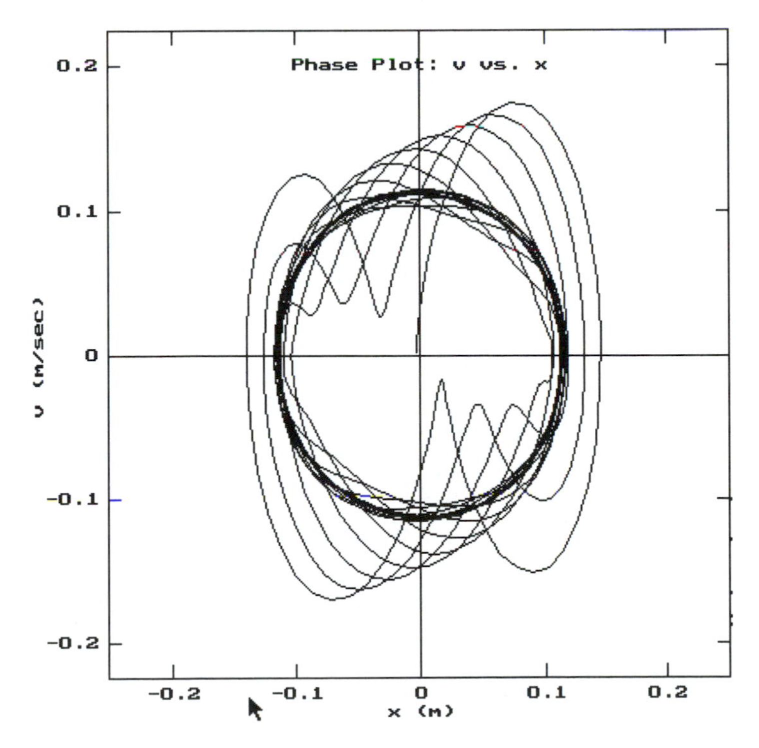

Figure 3.2: Driven pendulum started from rest at the origin and settling into a limit cycle.

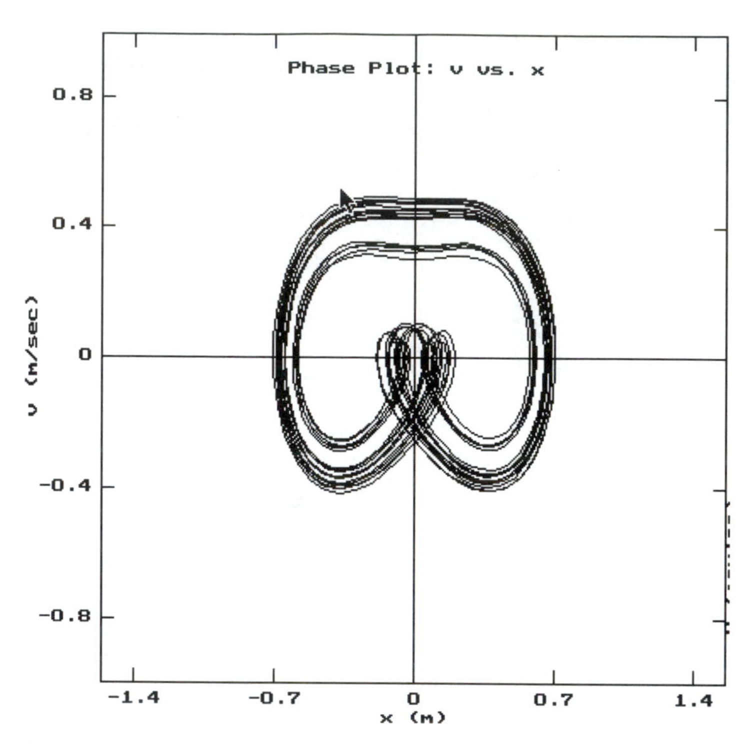

Figure 3.3: Phase diagram of a system with a long period.

3.2.6 Strange Attractors

Periodic attractors can have one loop in the phase diagram, as the clock pendulum does, or many loops and still be periodic. Other motions have no periodic structure and appear completely random until one looks at their Poincaré diagram. These are the *strange attractors*, and they are much easier to demonstrate than describe. You can find an example in the tutorial section Introduction to Poincaré Maps.

Figure 3.3 shows a periodic attractor with a long period, while Figure 3.4 shows a chaotic attractor.

We can tell that Figure 3.3 really is periodic by looking at the corresponding Poincaré diagram, Figure 3.5. In contrast, Figure 3.6 shows the Poincaré diagram of a chaotic system, in this case a pendulum.

Chaos

The presence of a strange attractor is an indicator of chaos, which is more precisely defined as existing when two trajectories with nearly the same initial conditions diverge from each other so rapidly that the distance between them grows exponentially with time. An excellent short introduction to chaos was provided by Max Dresden in *The Physics Teacher*.[13] A longer introduction may be found in Gleick's book.[14] Recent work on chaos shows that such a system can often be guided into a periodic orbit by small control impulses.[15,16]

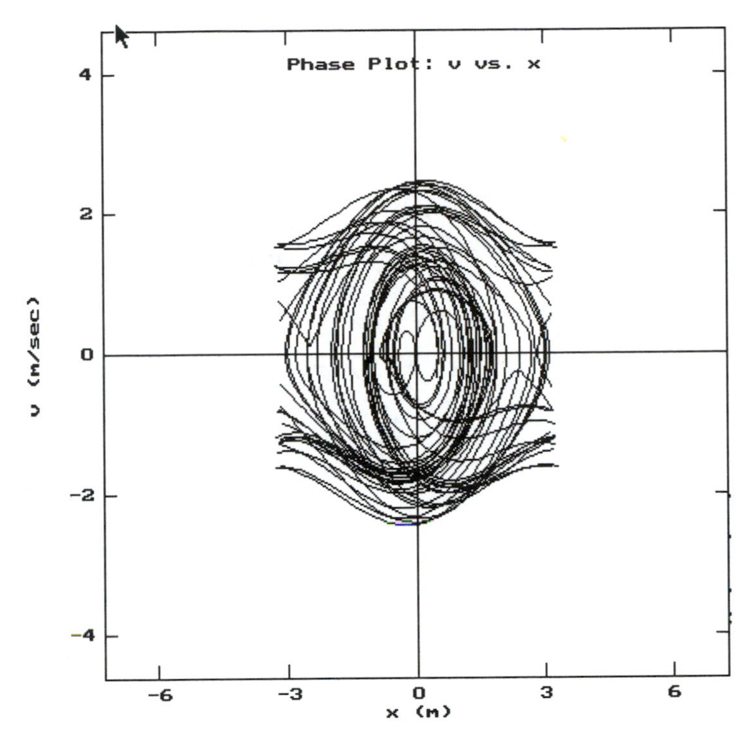

Figure 3.4: Phase diagram of a pendulum moving chaotically.

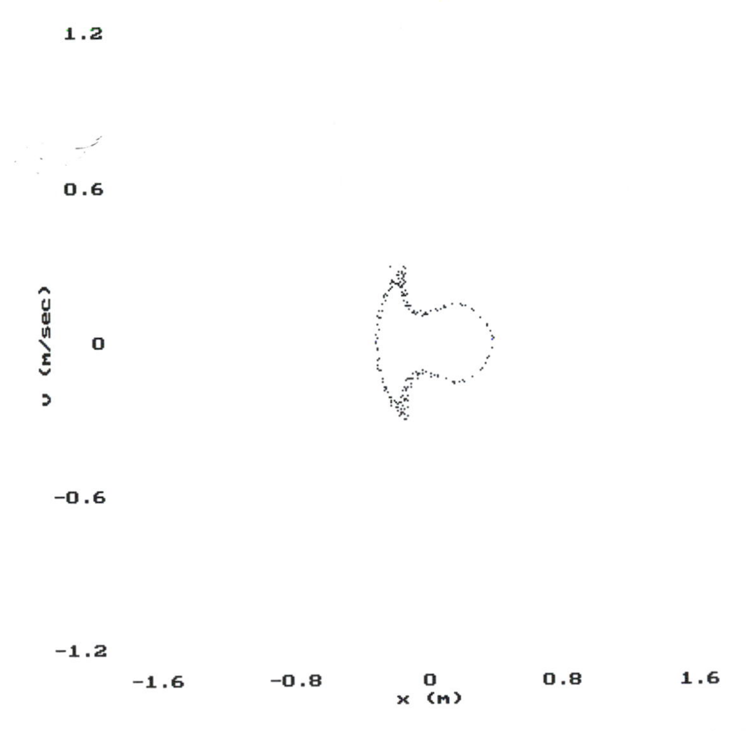

Figure 3.5: Poincaré diagram of the same long period motion shown in Fig. 3.3.

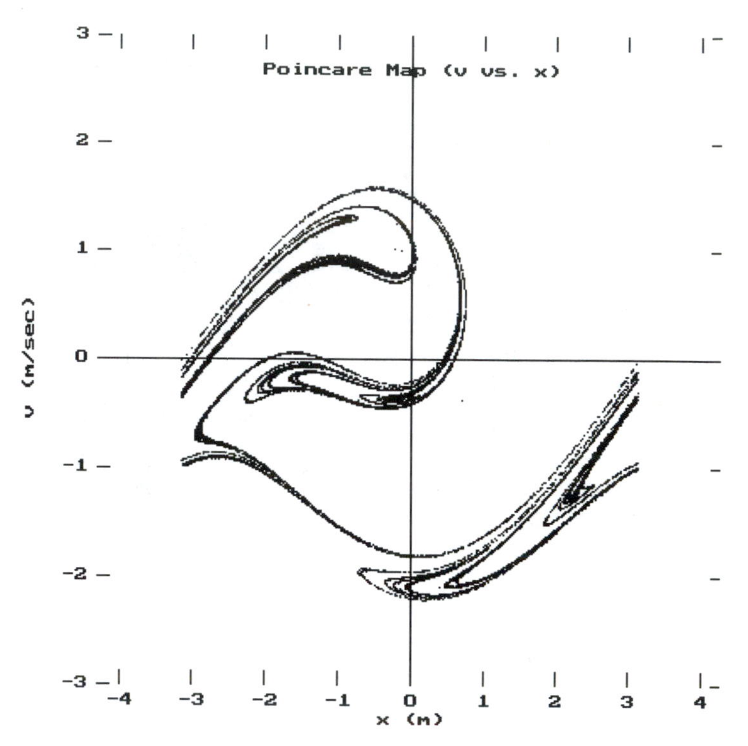

Figure 3.6: Poincaré diagram of the pendulum in chaotic motion.

3.2.7 Computational Approach

The simulation described here solves these one-dimensional equations using Runge-Kutta-Fehlberg[20,21] fourth-order numerical integration. For each model, one procedure calculates the force and another calculates the potential energy. For two systems, procedures are also provided to calculate polynomial approximations to the exact equations; only the first two non-trivial terms in the polynomial expansion are used.

3.3 Types of Oscillators

Four of the systems available in the simulation program are described by polynomial approximations to real systems. These are traditionally chosen for textbook discussion because of their mathematical tractability. Two others are numerical simulations of real physical systems. The approximations are the harmonic oscillator (polynomial force of degree 1), the symmetrical cubic oscillator (polynomial force of degree 3), the asymmetric oscillator, and the Van Der Pol oscillator (force dependent on x and x^2v). The physical systems are the pendulum and a mass on a spring stretched between two fixed points. The separation of the fixed points is adjustable, and the motion is confined to be perpendicular to the line joining the fixed points. Even here, the restriction to one-dimensional motion is an approximation. Removing it would get us into the domain of coupled oscillations, addressed in chapter 7. The computer simulation is again an approximation, replacing continuous motion with discrete time steps.

Analytical treatments of some of the models appear in textbooks[1-12] to varying degrees of completeness.

3.3.1 Simple Harmonic Oscillator

The simple harmonic oscillator is ubiquitous in textbooks because a complete analytical solution is possible. It is included here for comparison purposes. The equation used is

$$F = -kx - Dv + F_0 \sin \omega t. \tag{3.1}$$

Here, k is the spring constant; the second term is a damping force proportional to velocity; the last term is a sinusoidal driving force with angular frequency ω. The last two terms appear in all the models, but the default values for D and F_0 are zero, so that the default motion is undriven and undamped.

3.3.2 Asymmetric Oscillator

This is meant to illustrate a common situation in atomic and molecular physics where the two sides of the potential well have different shapes:

$$F = -kx - 3A_2x^2 - 4A_3x^3 - Dv + F_0 \sin \omega t. \tag{3.2}$$

Most of the constants have the same meaning as in Eq. 3.2. A_2 controls the asymmetry, and the third power term controlled by A_3 is necessary because otherwise the potential minimum would only be a local minimum. The x^2 term eventually overwhelms the linear term, turning the force on one side into a repulsive force. The x^3 term is necessary to keep the force at infinity attractive on both sides of equilibrium. You can demonstrate this with the simulation by setting the coefficient A_3 to zero. The period of this oscillator not only varies with amplitude, but higher harmonics appear in the frequency analysis.

3.3.3 Cubic Oscillator (Duffing's Oscillator)

$$F = -kx - C_3x^3 - Dv + F_0 \sin \omega t. \tag{3.3}$$

C_3 is simply a coefficient whose physical meaning depends upon the system that is being approximated by the equation. The symmetric cubic oscillator is often used as an illustration of approximation techniques and has a number of interesting features. It can either be hard or soft, depending on the signs of the coefficients k and C_3. Its period varies with amplitude, and it provides an example of a bent resonance curve with hysteresis and two basins of attraction. It is also a well-studied chaotic system. Khosropour and Millet[17] describe a simple apparatus to demonstrate this system. Baierlein[1] (section 3.2), among other textbooks, gives an approximate description of this system. Olson and Olsson[18] discuss the chaotic behavior of the system.

3.3.4 Van Der Pol Oscillator

$$F = -kx + D(X_0^2 - x^2)v + F_0 \sin \omega t . \tag{3.4}$$

The Van Der Pol oscillator exhibits both a limit cycle and chaos. It was an early model for electronic oscillators, whose nature depends on the presence of a limit cycle. Baierlein[1] (section 3.3), among other texts, gives an approximate description of this system.

The undriven ($F_0 = 0$) system with $D > 0$ loses energy when $x > X_0$ and *gains* energy when $x < x_0$. The equation represents a simple approximation to the behavior of electrical oscillators used in signal generators, TV and radio transmitters, and the like; where the energy gain comes from a battery or power supply.

Given what we know today about the complexity of the behavior of equations such as Eq. 3.4, it was a triumph of early electrical engineering to make such circuits well behaved with what was known at the time. Such systems sustain oscillation even without a driving force F_0, and this is the purpose for which they are used. The energy comes not from a driving force term but from the non-linearity acting upon DC currents in the real physical system.

3.3.5 Pendulum

$$F = -\frac{g}{L} \sin \theta - Dv + F_0 \sin \omega t . \tag{3.5}$$

The parameters are g, the acceleration of gravity; and L, the length of the pendulum.

The pendulum is of considerable practical and historical engineering importance, the grandfather clock being only the most familiar example. It is a soft oscillator, with the period increasing with amplitude, a constant source of vexation to clock makers and a stimulus to their ingenuity in finding ways to keep the amplitude constant. It is also a beautiful example of chaotic motion and basins of attraction to both periodic and chaotic attractors. Perman and Hamilton[19] discuss the erratic transient behavior of the undamped pendulum.

3.3.6 Springs Between Two Walls

Two springs are stretched between two fixed walls as shown in Figure 3.7 with a mass in the center. The mass is constrained to move only in one horizontal dimension:

$$F = -2k(L - L_0)x/L - Dv + F_0 \sin \omega t , \tag{3.6}$$

where

$$L^2 = x^2 + (W/2)^2 . \tag{3.7}$$

The spring constant of each spring is k, and L_0 is the unstretched length of the spring, L is its stretched length, and W is the distance between the walls. This system is approximated, but very poorly, by the cubic equation, whose parameters k and C_3 change and even change sign as the wall separation is changed. Approximate treatments of this system are discussed extensively by Baierlein[1] (section 3.1).

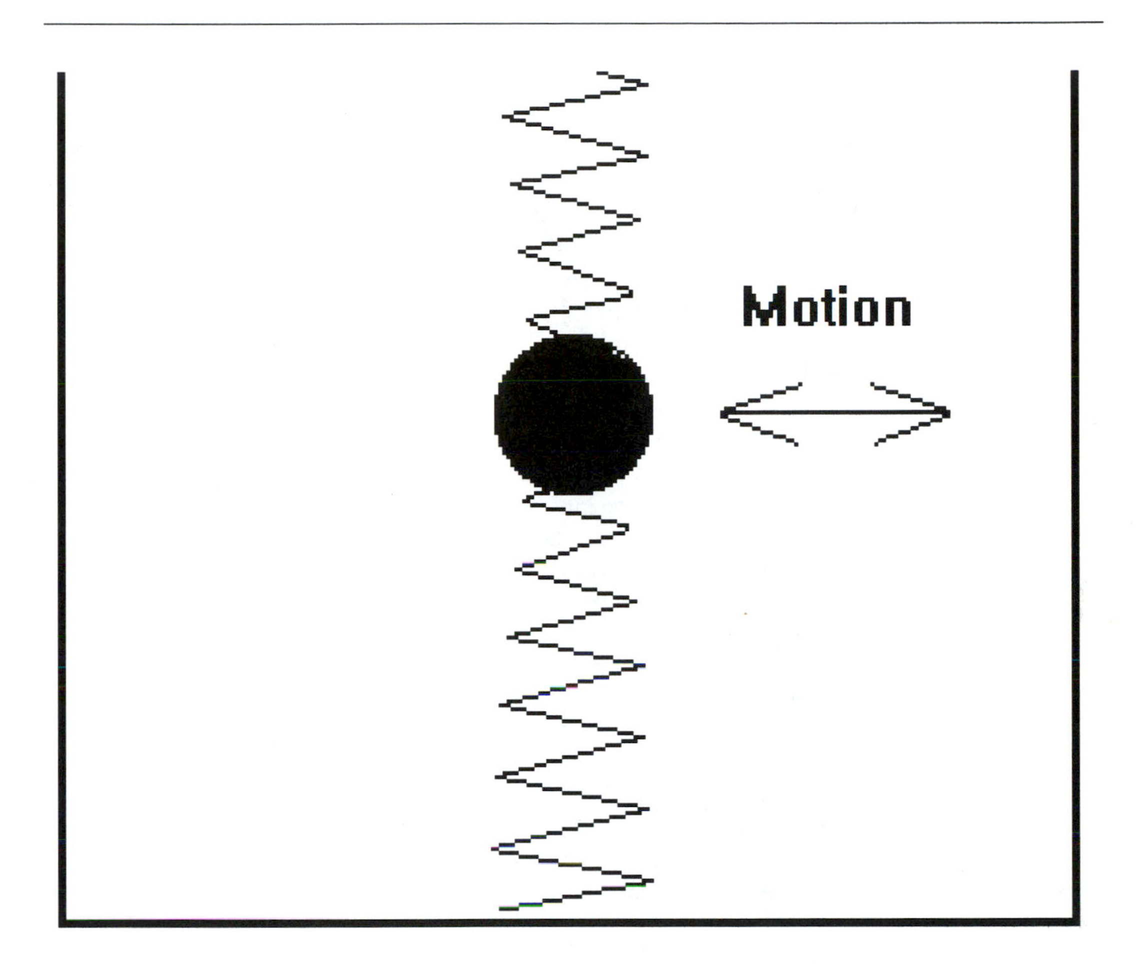

Figure 3.7: A spring between two walls.

This system of two springs holding a mass stretched between fixed points is described by the same equation as an air-track car attached by a spring to a point on the ceiling. Then gravity performs the function of keeping the motion one-dimensional, making the second spring superfluous. In both cases, springs tend to buckle when compressed, making it difficult to realize all features of the motion in the lab. It is a hard system, the restoring force increasing faster than linear. This system exhibits chaos.

Besides providing a real system to which the cubic equation is an approximation, this system is an example of a phenomenon called spontaneous symmetry breaking, of considerable importance in several areas of physics such as elementary particle physics. Spontaneous symmetry breaking occurs whenever an initially symmetrical system becomes asymmetric, though the change of a parameter which is not in itself asymmetric.

In this case, the system is clearly a symmetric one: there is no difference between motion to the right and motion to the left. However, if you push the two walls close enough together, the springs are compressed instead of stretched. There is nothing to prevent them from pushing the mass to one side or the other, and they do. The wall separation is not an asymmetrical parameter, and yet the system is no longer symmetrical.

We say that the symmetry has been broken, and we call it spontaneous symmetry breaking because it was not done by introducing anything asymmetrical. Rather it has happened by itself as a result of a change in a parameter, the distance between the walls. There are now two equilibrium positions instead of one. The potential well acquires a hump in the middle.

3.4 Using the Program

3.4.1 Setting a System Into Motion

A motion can be started in three ways: by clicking the mouse button with the cursor in the phase diagram or in the potential diagram, or by pressing the **Enter** key.

Click in the Phase Diagram. The position of the cursor selects the initial x and v. A line is drawn in the potential diagram at the corresponding energy.

Click in the Potential Diagram. The vertical position of the cursor determines the energy of the motion, which is always started with $v = 0$. x will be either positive or negative depending on the horizontal position of the cursor. A line is drawn across the potential diagram at the energy selected. If the motion is driven, damped, or both, then the energy selected is the initial energy, and the energy line drawn does not reflect the subsequent changes in the energy.

Press Enter. This option is provided for those so unfortunate as not to have a mouse. The energy and some other parameters can be changed in steps by the use of F keys as indicated at the bottom of the screen; their use is discusseed further in chapter 1. Parameter values between the steps can be obtained by selecting **Drive, Damping & Parameters** from the **Forces** menu.

3.4.2 Graphical Display

The user sees a screen divided into several parts, with some choice of what is displayed in each. In the upper right is a picture of the system. This can be animated to show the actual motion when desired. Since this animation also slows the calculation, it can be turned off.

Below the system picture is a potential energy diagram for the system. When animation is turned on, a ball rolls up and down the potential well. One of the ways of starting the system is to move the cursor into this diagram and click the mouse button.

The larger left-hand part of the screen can be divided into one, two, or four windows, as the user chooses. These windows display the user's choice of graphs of position or velocity against time, velocity against position (phase diagram), a Poincaré diagram, or a return diagram which plots against each other velocities calculated at successive time intervals of one driving period.

3.4.3 Changing Parameter Values

All relevant parameters—including initial values, spring constants, friction, and driving force parameters—can be changed by selecting **Drive, Damping & Parameters** from the **Forces** menu.

Parameters relevant to the display, such as graph scales, can be changed by selecting **Display Options** from the **Display** menu.

Changing the Wall Separation. The **Two Walls** system behavior changes drastically as the distance between the walls changes. You can change it with the mouse simply by dragging the upper wall or by placing the cursor where you want the upper wall and clicking the mouse button.

3.4.4 Program Menus

This section describes the menu options found in the program. They may be selected with the mouse, or by pressing F10, **arrow** keys, and the **Enter** key.

- **Files**: Allows access to special files that can be created and read by the program, information about the program, program configuration, and program exit.
  - **About Anharm**: Brief description.
  - **About CUPS**: Brief description of the project.
  - **Configure**: Allows you to change the path to the directory in which parameter and Poincaré map files are stored (shortcut key Alt-d), change the colors (shortcut key Alt-c), or check the amount of memory available to the program (shortcut key Alt-m). You can change the colors to black-and-white to print the screen on a black-and-white printer.
  - **Save Parameters**: When you have found a particular set of parameters that you might want to use again, you can save them with this option. You will be prompted for a file name.
  - **Read Parameters**: Allows you to recover parameters that have been saved with the previous option. You are allowed to select from a list of the files available. All parameters (except initial conditions), including the force type, are set to the values saved.
  - **Save Poincaré Maps**: After you have constructed a Poincaré diagram, which can take considerable time, you can save the whole Poincaré movie with this command. You will be prompted for a filename.
  - **Read Poincaré Maps**: This will present you with a list of filenames of previously saved Poincaré movies. The selected one will be read in and immediately displayed.
  - **Compare Two P Maps**: This will read in any two previously saved Poincaré movies and display them in two windows on the screen.
  - **Read Search file**: This reads in a file created by **Search Parameter Space**. It will consist of a set of single Poincaré diagrams, which will be reviewed for you on the screen. Each diagram will have been made with a different set of parameters. Arrow and letter key options listed at the top of the screen allow you to quit, pause to inspect any diagram, move backward (over a limited range) or forward in the file, and extend any diagram by starting a computation with the corresponding parameters. You can save particularly interesting diagrams with their parameters to a new file whose filename will be derived from the name of the file you are examining.

- **Save History**: Save the history of the most recently executed motion in either of two forms, both ASCII text files. One form lists t, x, and v in three columns with a header, including an identifying description that you provide. This can be used for any analysis you desire. The other form is simply a list of x-values with no header, the form required by *Chaos Dynamic Analyzer* from Academic Software.
- **Exit Program:** Can also be accomplished by pressing the shortcut key **Alt-x**.

- **Choices**: This menu contains choices to clear the screen, turn animation on and off, make special plots, and the like.
  - **Clear**: The graphs do not automatically clear when you start a new motion with different initial conditions. This allows you to compare different motions. When the screen gets too cluttered, use this command.
  - **Clear (& Rescale)**: This adjusts the scale of the graphs so the most recent motion fills the window.
  - **Plot in Both [One] Wells**: Some potentials have two wells. If the energy is below the top of the partition between the wells, the motion is confined to the well in which it starts. This menu item allows you to choose to plot in both wells each time you start the motion.
  - **Animate [Off]**: This turns on and off animation, which displays the actual motion of the system in the upper right window and indicates its current position in the phase diagram. Turning animation off considerably speeds up the calculation.
  - **Change Potential Range**: This allows you to change the scale of the potential diagram to a value you select, if the automatic scale changes do not do what you want. You will be asked for a new value for the full scale energy. (The scale changes automatically when you choose an energy in the top or bottom 10% of the range on the diagram.)
  - **Plot Period**: This calculates and displays a graph of the period of the system as a function of energy for (usually) five different values of a system parameter such as spring constant or gravitational acceleration. Once it has been calculated, it can be displayed again quickly, from the **Display** menu, as long as you have not changed the force type. The graph takes the place of the system diagram in the upper right of the screen until construction is finished, when it moves to fill the left hand side of the screen.
  - **Plot Resonance**: This plots resonance curves for two to five different values of a system parameter. This takes a long time, since the transient must be allowed to die away at each frequency. An input screen allows you to change parameters and choose whether to search for a repulsive branch (if present) or to approximate it. This branch, the overhanging part of a resonance curve that is bent to one side, is present only when there are two different stable motions at the same frequency, as occurs for some oscillators. The search option is *really* slow. This graph also can be displayed again until the force is changed.

- **Frequency Analysis**: Displays a discrete Fourier[24] transform of the most recent motion. Several choices are offered:
  * **Magnitude and Phase**: Plots the magnitude of the transform in yellow and the phase in green. The number of points in the history is displayed. You can zoom in (repeatedly) by a factor of two, expanding the frequency scale of the graph, and back out again.
  * **Real+Imaginary**: Display the real part of the transform in yellow and the imaginary part in green. The zoom option is again available.
  * **Windowed**: Multiplies the motion history by one of three window functions.[25] Since the function analyzed starts and stops abruptly, artifacts result. Windowing alters the artifacts, as discussed in the reference.
  * **Test Function**: Plots the transform of either a sine, square, or triangle wave. You can select the period, number of points, and phase shift as well as any of the three windowing functions.
- **Poincaré Map Movie**: Only available after a Poincaré diagram has been displayed. The movie shows a sequence of Poincaŕe diagrams taken at different phases of the driving force. The F keys allow you to slow or speed the motion, pause it, and end it. F2 truncates each frame by removing the first few points, thus removing initial transients. It can be pressed repeatedly to remove more points.
- **Return Map Movie**: Activates a similar movie of return diagrams.

• **Display**: This menu selects the graphs displayed, and allows you to set their scales as well as a number of other options. The diagram of the system and the potential diagram are always displayed.

- **x vs t**: Displays as a large single graph.
- **v vs t**: Displays as a large single graph.
- **v vs x (Phase Diagram)**: Displays as a large single graph.
- **Poincaré Map**: Displays as a large single graph.
- **Return Map**: Displays as a large single graph.
- **Display Two Graphs**: Display two selected graphs one above the other. Use F keys to select the two graphs.
- **Display Four Graphs**: Divides the left side of the screen into four windows, displays x vs. t, v vs. x, the Poincaré diagram, and either the v vs. t graph or the return diagram. Use F keys to select one of the latter two.
- **Display Period Graph**: Displays a previously calculated period graph.
- **Display Resonance Graph**: Displays a previously calculated resonance graph.
- **Display Options**: Allows you to change the following:

* **Animation Delay Time**: Use to speed up or slow down animation.

* **Undriven Time Step**: The basic computation time interval for undriven motions.

* **Driven Time Step**: The basic computation time interval for driven motions as a fraction of the period of the driving sine wave.

* **Graph Time Scale**: The length of the time axis on x vs. t and v vs. t graphs. Note that for times larger than this the graph wraps around but the axis labels are not changed.

* **Graph *X* Scale:** Change if the graph is too large or too small.

* **Graph *V* Scale:** Change if the graph is too large or too small.

* **Number of Movie Frames**: Poincaré movies are sets of maps made at this many equally spaced intervals in the phase of the sinusoidal driving force. Take more if the movie jumps too far between frames. If you ask for more than the available memory allows, the number will be automatically adjusted. Twenty-five is about as many as you can get unless the program is specially compiled to use extended memory; this number depends on the exact set-up of your computer. Thirty-two is the largest number the program allows.

* **Stop After Two Periods**: This is adjusted automatically by the program, but you may sometimes prefer to override the program's choice. **Yes** means that the motion will stop automatically; **No** means that the motion will continue until you click the mouse or press **Enter**.

* **Write Over Time Graphs**: Controls what happens when the time graph wraps around. **Yes** means that the older graph will be preserved for comparison purposes. The graph changes color periodically to help you distinguish the several older graphs. **No** sets the graph to erase the old one just ahead of the new one.

* **Mark Poincaré in Phase**: An aid to understanding the Poincaré diagram. When turned on, each point in the Poincaré diagram will be marked in the phase diagram as well (both diagrams must be displayed). This option dramatically slows the calculation.

* **Demonstrate Phase Plot**: Displays the v vs. t graph with axes reversed beside the phase plot and the x vs. t graph beneath it, so that the respective x- and v-axes correspond, thus showing the two time graphs as projections from the phase graph.

- **Forces**: Here is where you select among the different force models, along with some related options. The force models are all at the bottom of the menu.

  - **Asymmetric**: $F = -kx - 3A_2x^2 - 4A_3x^3 - Dv + F_0 \sin \omega t$. An approximation to inter-atomic forces in molecules.

  - **Cubic (Duffing)**: $F = -kx - C_3x^3 - Dv + F_0 \sin \omega t$. This equation has been extensively studied by Ueda[26] and others. Thompson and Stewart[27] have a summary of this discussion. The quantities k and C_3 can have either sign, or be zero. Positive C_3 represents a "hard" spring that gets stiffer as it is stretched; negative C_3 represents a "soft" spring that gets weaker as it is stretched. Negative k and positive C_3 result in a double potential well like that of the spring stretched between two walls, to which this is an approximation.

- **Pendulum**: $F = -(g/L)\sin x - Dv + F_0 \sin \omega t$. This familiar system is capable of bizarre behavior when it is driven hard enough to swing over the top. The adjustable parameters are g, the acceleration of gravity; and L, the length of the pendulum.

- **Simple Harmonic**: $F = -kx - Dv + F_0 \sin \omega t$. This linear system is included for the sake of completeness and comparison.

- **Two Walls**: $F = -2k(L - L_0)x/L - Dv + F_0 \sin \omega t$, where $L^2 = x^2 + (W/2)^2$; k is the spring constant of the spring connecting the body to one wall, L_0 is the unstretched length of that spring, L is its stretched length, and W is the distance between the walls. It is only approximated by the cubic equation: the parameters k and C_3 of the approximation change sign and magnitude as the wall separation is changed.

- **Van Der Pol Oscillator**: $F = -kx + D(X_0^2 - x^2)v + F_0 \sin \omega t$. Here we have energy loss when $x > X_0$ and *energy gain* when $x < X_0$. X_0 therefore sets the size of the limit cycle.

- **Explain Current Force**: Displays information about the model now being calculated, including its equation and its parameters.

- **Drive, Damping & Parameters**: Allows you to type in exact values for wall separation, pendulum length L, Van Der Pol limit cycle size X_0, spring constant k, mass m, cubic coefficient C_3, driving force F_0, angular frequency ω, drag coefficient D, asymmetric oscillator constants A_2 and A_3, approximate force coefficients A_1 and A_3, and initial position and velocity.

- **Interesting Parameters**: Displays a new menu for each model. Each one has a number of items of interest for that model, always including undriven, undamped, and the default parameter values.

- **Search Parameter Space**: This allows you to discover interesting parameter values for yourself. Since this can take a long time, it is designed to run unattended, writing results to a file to be examined later. That file will contain a 30-point Poincaré map for each set of parameters investigated. Menu item **Read Search Map** on the **Files** menu allows review of the results and selection of set of parameter values of interest.

  You will be asked for a file name, and then to type in values for the parameters you wish to start with, how many changes to make in each, and how much to increment each parameter by.

  After the first map has been calculated, you are provided with a time estimate for the search. You can quit at any time, or truncate the individual maps if the results are obviously uninteresting.

- **Series Approximation**: Switches to a series approximation:

  $F = -A_1 mx - A_3 mx^3 - Dv + F_0 \sin \omega t$ (**Two Walls**) or
  $F = -(g/L)\theta + \theta^3/3! - Dv + F_0 \sin \omega t$ (**Pendulum**).

  The potential diagram displays both the approximate and exact potentials, and the graphs are not cleared, so you can compare the approximation to the exact solution with identical initial conditions.

- HELP!!

  - **Help**: Changes the menu into a **Help** menu. Selection of any menu item then
    displays one or more screens of information about that item, and this can be done repeatedly. Leave the help system by clicking the mouse anywhere but in a menu, pressing **Enter**, or selecting **Quit Help**.
  - **Tutorial**: Allows you to enter the tutorial at any time.

3.5 Exercises

3.1 **Comparing Periods**

(a) Does the period of the pendulum increase or decrease as the amplitude of the motion increases?

(b) Does the period of the **Two Walls** system with unbroken symmetry (a single potential well, walls far apart) increase or decrease as the amplitude of the motion increases?

(c) Does the period of the **Two Walls** system with *broken* symmetry (double potential well, walls closer together) increase or decrease as the amplitude of the motion increases? (Answer for both cases: motion with energy above and below the hump in the potential well [Figure 3.8].)

(d) What happens to the period as the energy gets very near the hump?

(e) Explain each of these behaviors in terms of the forces acting on the relevant system.

(f) Relate these period changes to particular features of the x vs. t, v vs. t, and v vs. x graphs for each case.

3.2 **Period Peaks**
With the **Two Walls** force in effect, select **Plot Period** from the **Choices** menu.

(a) What significant event(s) correlate(s) with the peaks in the period vs. energy? What happens to relevant parameters when these events occur? (Turning animation on may help, but it takes a long time, so don't do it the first time around or unless you need to.)

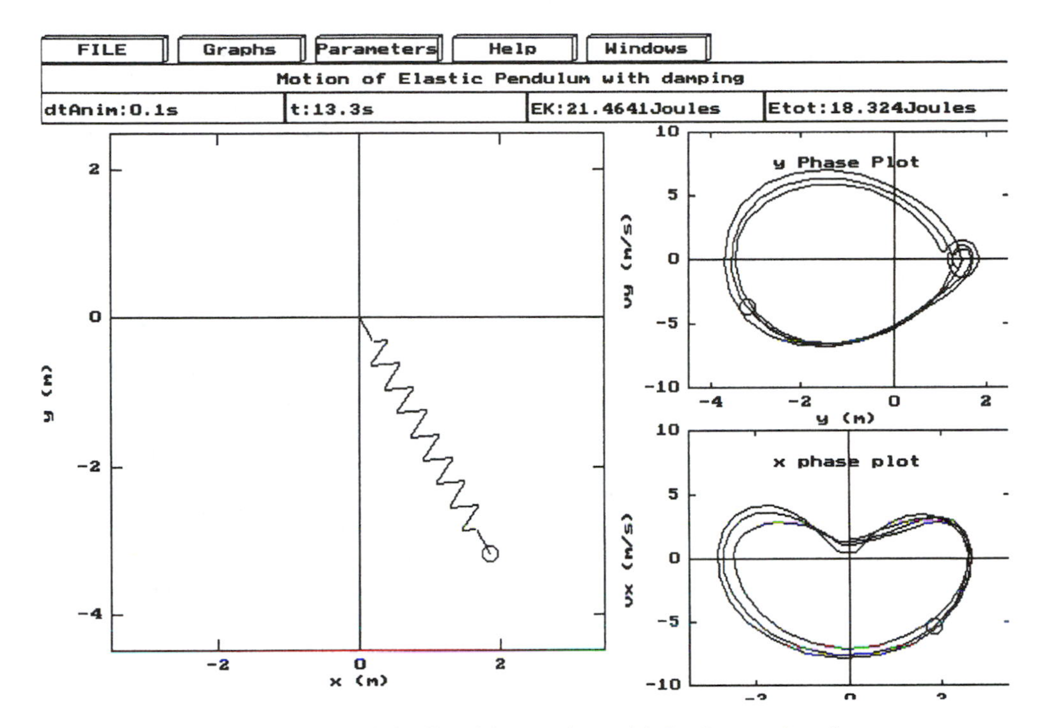

Figure 3.8: Consider motion with both energies shown.

(b) Answer the same questions for the pendulum period vs. energy graph.

(c) How do the relationships explain the existence of the peaks? Give detailed descriptions of the motion at the peaks as compared to elsewhere.

3.3 Speed of Approach to a Limit Cycle

(a) Start the Van Der Pol oscillator by clicking the mouse button with the cursor at the origin of the phase diagram. How many periods does it take for the motion to merge with the limit cycle? (Estimate to a tenth of a period, if you can.) How many periods does it take starting from each of the four corners of the phase diagram and from various other places around the outside edge of the diagram?

(b) What regularity do you observe about motions starting from various places around the edge?

(c) Pick several points on the trajectories in the phase diagram and explain how the force given by Eq. 3.4 predicts that the system will evolve from each of those points in just the way that it does: why does the trajectory go on from there in the direction that it does? Use the insight you have gained from that to explain the regularity referred to in b.

(d) Select **Drive, Damping & Parameters** from the **Forces** menu; note the present value of the **Drag Coefficient**, and change it to 0.3. Answer the questions in part a. Make a neat table of the numbers of periods from a and from d to facilitate comparison. State in words a generality about the changes. Have they all changed by comparable amounts? (Drag coefficient is D in Eq. 3.4. That isn't really a good name for it here, but we are using the same entry table for several force laws.)

(e) Explain how the force given by Eq. 3.4 predicts the results of your generalization in d.

(f) Change the **Drag Coefficient** to 0.2 and enter values in your table corresponding to those you entered from a and d. Does the generalization still hold? Describe anything new that has appeared.

(g) What is the overall effect of reducing the **Drag Coefficient**, and how does Eq. 3.4 predict that?

3.4 **Improving Series Approximation** (Programming Project)
Add more terms to either of the series approximations; some hints are given in Exercise 15. The relevant functions have names beginning in "Approx." Calculate what the values of the additional coefficients should be by expanding the exact force laws in a power series. (This is a major task for the **Two Walls** system.)

3.5 **Understanding Fourier Analysis**
Select **Frequency Analysis, Test Function**.

(a) How do the analyses of the three different waveforms differ? Do the results agree with what you would expect?

(b) How do the results change if you use 40 points instead of 512?

(c) What is the effect of windowing on the 160-point sine wave? Is there any difference between the three windows? Repeat for the 160-point triangle wave.

3.6 **Basins of Attraction**
Select **Two Walls** with a wall separation of 0.8, $F_0 = 0.3$, $\omega = 0.45$, and **Drag Coefficient** = 0.1.

(a) Start the motion by clicking with the cursor at -1.0, 0.0 and wait until the motion reaches a limit cycle. Start it again at $+1.0$, 0.0 and wait until it reaches a limit cycle. Describe the difference between the two limit cycles. Assign a unique letter or color to each. (Be sure that the scale of the phase map is the default: v should run from -0.6 to 0.6, and the labeled ticks for x from -1 to 1.)

(b) Explore the x-v plane to find the boundary between the basins of attraction of the two limit cycles: start the motion at some point and on a paper graph of the plane write at the starting point the letter or make a colored dot corresponding to the limit cycle reached. A useful procedure is to start with the origin, the ends of the axes, and the corners of the graph. Then look halfway between any two points that belong to different basins (different limit cycles). Extend this process into the quadrants through which the boundary runs to find out whether it is straight or curved.

3.7 **More Basins** (More Work and More Credit)
Use the parameters of Exercise 6 and change the wall separation to 0.4 and repeat the process. The boundary is much more complex this time, and you will have to look carefully to find all the pieces. Is it singly connected or multiply connected?

3.8 **And More Basins** (Still More Work and Still More Credit)
Repeat Exercise 7 with the Drag Coefficient changed to 0.2. The boundary is *very* complex.

3.9 **Still More Basins** (Cooperative Class Project)
Explore what happens to the basin boundary(ies) when you vary any of the parameters (**Wall Separation**, F_0, ω, **Drag Coefficient**). You might want to do some quick exploration to find out which are more interesting before you settle on a final choice.

3.10 **Many Basins & Poincaré Map**
Select **Long Period** from the **Two Walls** menu of the **Interesting Parameters** choice on the **Forces** menu. Choose **Display Two Graphs** from the **Display** menu and press F4 and F5 to select the Phase Map and the Poincaré Map.

(a) How many different long period attractors can you find? The Poincaré diagram will help you tell them apart.

(b) Explore some of the basins of attraction. You may well find that the basins are multiply connected and very fine in structure. Look for some of the fine structure rather than attempting to find all the boundaries, which would probably be an exhausting task. The procedure in Exercise 6 may be helpful.

3.11 **Systematic Exploration of Many Basins** (Programming Project)
Modify the **Search Parameter Space** routine—Procedure **SearchParams** in ANHMENU.PAS—to automate and systematize the search suggested by Exercise 10.

3.12 **Basins of Attraction in Other Systems** (Major Project)
Find examples of multiple basins with other force laws. Likely places to look are areas where the trajectory in the phase plane has loops.

3.13 **Unstable Equilibrium Search**
There is an unstable equilibrium point in the Van Der Pol phase plane. Find it and demonstrate its existence, as follows:

(a) Intuition method: Where do you think the system might be confused about which way to go? Analysis method: At what point does the equation say that the force would be zero?

(b) How do the trajectories against time differ when you start near the equilibrium point? What visible difference is there compared to starting anywhere else?

3.14 **Transient Beats**

Select the **Simple Harmonic** force. Adjust the **Drag Coefficient** so that motion decays to a small fraction (10% to 20%) of its original value in about ten periods. Then add a small driving force amplitude (0.05 is reasonable). Launch the motion by clicking at the origin of the phase diagram so that x and v both start from 0.

(a) What is special about the motion at $\omega = 1$?

(b) What happens as ω becomes more and more different from 1? There is a regularity; describe it.

(c) What happens if you launch the motion with x, v, or both different from 0?

(d) How does changing the **Drag Coefficient** change the motion?

(e) Do the results you just obtained hold for other force systems?

3.15 **Noise** (Programming Project)

Add noise to any of the driven force systems by changing the appropriate acceleration function. These are all found in ANHPHY.PAS, and the function names end in "Accel." Use the variable named **UserPar** to control the amount of noise; it appears on the **Drive, Damping & Parameters** entry screen as **User Parameter**. It will have a value of 1 until you change it.

How does the addition of noise change the motion? Investigate short period, long period, and chaotic motions.

It has been found[22] that identical noise applied to two different trajectories forces them to the same trajectory. You might wish to explore the possibility of this in the systems discussed here.

3.16 **Additional Force** (Programming Project)

Modify any of the force laws (see Exercise 15) to suit yourself. The variable **UserPar** is available to use as an additional constant parameter in your new force law.

3.17 **Transition to Chaos**

Change the display to show both the phase map and the Poincaré map. Use **Display Options** to set the **Number of Movie Frames** to 24. Select **Chaotic** from the **Two Walls** menu of the **Interesting Parameters** choice on the **Forces** menu. Then use the **Drive, Damping & Parameters** entry screen to change parameter values. When looking for a limit cycle, you can use the C key to clear both diagrams and continue running, thus removing the confusing initial transient. This does not clear points from the movie accessed from **Flip Poincaré Maps** on the **Choices** menu; that is done with the F2 key while viewing the movie.

(a) Run with the default **Chaotic** parameters to see what the chaotic evolution looks like. Then run the movie to see what that looks like. Change the **Drag Coefficient** to 0.5. You will get a limit cycle which evolves to a single point in the Poincaré diagram. (Use C to see that, F2 to see it in the movie.)

(b) Change the **Drag Coefficient** to 0.4. The Poincaré map will eventually evolve to just two points.

(c) Now explore near 0.4 to see how the system evolves from a limit cycle to chaos. You should be able to find at least one more doubling of the period to four eventual points in the Poincaré diagram. What else do you find? Take careful notes, and perhaps save the Poincaré maps using **Save Poincaré Maps** on the **Files** menu to review them later. Write a description of what you find, with sketched diagrams.

3.18 **Exploration of Chaos**
Start as in Exercise 17. Explore changing either F_0, ω, or the wall separation as follows:

(a) Find (if possible) a value of the parameter for which there is a single period limit cycle.

(b) Then find regions of the parameter in which interesting changes take place, and record them.

3.19 **Basins of Attraction and Chaos** (Project, Could Be Cooperative)
Proceed as in Exercise 18, but explore changing two or all three parameters.

First explore the axes along which one parameter changes, then explore systematically the effect of changing two or more parameters simultaneously.

3.20 **Fourier Analysis of Chaos**
Compare the frequency analyses of the chaotic motions of the pendulum and the **Two Walls** system.

(a) List the frequencies present in each case, with their magnitudes.

(b) Describe the similarities and differences between the two.

3.21 **Numerical Analysis and Chaos** (Major Programming Project)
It has been suggested that in some cases roundoff errors dominate chaotic systems and that in some model systems chaos is entirely the result of roundoff errors. It has also been found that different numerical analysis methods sometimes remove chaos from a model. Investigate this possibility for the models in the present simulation. The numerical analysis procedure **StepRK4** in the CUPS utilities is called twice in ANHPHY.PAS. Substitute a call to a different procedure written by you which you add to ANHPHY.PAS.

3.22 **Non-Sinusoidal Driving Force** (Programming Project)
Add higher harmonics to the driving force for any of the force laws. (See Exercise 15 for programming hints.) Describe how this changes the motion. Does it change it in the same way for different force laws?

References

1. Baierlein, R. *Newtonian Dynamics*. New York: McGraw-Hill, 1983.

2. Fowles, G. R. *Analytical Mechanics*. New York: Saunders College Publishing, 1986.

3. Goldstein, H. *Classical Mechanics*. Reading, MA: Addison-Wesley, 1980.

4. Marion, J. B. *Classical Dynamics of Particles and Systems*. New York: Academic Press, 1970.

5. Landau, L. D., Lifshitz, E. M. *Mechanics*. Reading, MA: Addison-Wesley, 1960.

6. Symon, K. R. *Mechanics*. Reading, MA: Addison-Wesley, 1971.

7. Slater, J. C., Frank, N. H. *Mechanics*. New York: McGraw-Hill, 1947.

8. Kittel, C., Knight, W. D., Ruderman, M. A. *Mechanics*. New York: McGraw-Hill, 1965.

9. Becker, R. A. *Introduction to Theoretical Mechanics*. New York: McGraw-Hill, 1954.

10. Taylor, E. F. *Introductory Mechanics*. New York: Wiley, 1963.

11. French, A. P. *Newtonian Mechanics*. New York: W.W. Norton, 1971.

12. Synge, J. L., Griffith, B. A. *Principles of Mechanics*. New York: McGraw-Hill, 1959.

13. Dresden, M. Chaos: A new scientific phenomenon—or Science by public relations? The Physics Teacher **30:**10, 74, 1992.

14. Gleick, J. *Chaos: Making a New Science*. New York: Viking, 1987.

15. Gang, H., Kaifen, H. Controlling chaos in systems described by partial differential equations. Physical Review Letters **71:**3794, 1993.

16. Chacon, R., Diaz Bejarano, J. Routes to suppressing chaos by weak periodic perturbation. Physical Review Letters **71:**3103, 1993.

17. Khrosropour, R., Millet, P. Demonstrating the bent tuning curve. American Journal of Physics **60:**429, 1992.

18. Olson, C. L., Olsson, M. G. Dynamical symmetry breaking and chaos in Duffing's equation. American Journal of Physics **59:**907, 1991.

19. Permann, D., Hamilton, I. Self-similar and erratic transient dynamics for the linearly damped simple pendulum. American Journal of Physics **60:**442, 1992.

20. Press, W. H. et al. *Numerical Recipes*. Cambridge: Cambridge University Press, 1986, p. 550ff.

21. Maron, M. J. *Numerical Analysis*. New York: Macmillan, p. 344ff, 1982.

22. Maritan, A., Banavar, J. R. Chaos, noise, and synchronization. Physical Review Letters **72:**1451, 1994.

23. Ablowitz, M. J., Schober, C., Herbst, B. M. Numerical chaos, roundoff errors, and homoclinic manifolds. Physical Review Letters **71:**2683, 1993.

24. Press, *op. cit.*, p. 381ff.

25. Press, *op. cit.*, p. 423ff.

26. Ueda, Y. In *New Approaches to Nonlinear Problems in Dynamics*. ed. P. J. Holmes, p. 311ff. Philadelphia: SIAM, 1980.

27. Thompson, J. M. T., Stewart, H. B. *Nonlinear Dynamics and Chaos*. New York: John Wiley and Sons, p. 6ff, 102, 1986.

4

Central Force Orbits

Bruce Hawkins

Then felt I Like some watcher of the skies
When a new planet swims into his ken

John Keats, *On First Looking into Chapman's Homer*

4.1 Introduction

A quick introduction to the features of the computer program is in the ORBITER walk-through in Appendix A.

Fascination with the motion of objects that came to be called "planets" extends back beyond the beginning of recorded history. Those who watched the starry skies paid them a good deal of attention and discovered that a few of the stars moved in relation to the others. The very word *planet* meant "wanderer" in Greek. The term is especially apt, because the motion is not steady and regular like that of the Sun, Moon, and stars, but turns backward and forward in the sky.

Attempts to understand these motions started with the simplest: the stars moved in circles about the Earth, all at the same rate. The Sun and Moon also moved in circles, but at a different rate, the Moon much faster. Since circles worked fine for all these bodies, people tried to use circles for the planets as well.

To accommodate the back-and-forth motion of the planets (Fig. 4.1), it was concluded that each planet must move on a small circle whose center moved along a larger circle centered on the Earth. Alas, fitting the observations required a complicated apparatus of circles with their centers moving in odd places. Changing to a sun-centered system of circles does not help this much, either. The observations could be matched, but the apparatus was undesirably cumbersome.

Eventually Kepler discovered that ellipses centered on the Sun fit the observations with a reasonably simple rule for the changing speed of the planets. To see how that explained the back-and-forth motion, pick **Retrograde Motion** from the

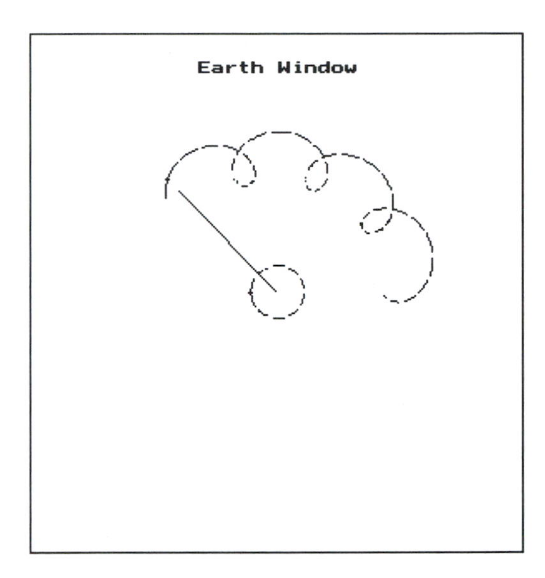

Figure 4.1: Retrograde motion (Jupiter and Sun shown) for four years, seen from Earth.

Systems menu. The planets move faster when nearer to the Sun, according to Kepler's law of areas, now understood as conservation of angular momentum, a direct result of the central nature of the force.

Increasingly accurate observation made possible by the advent of the telescope showed that even ellipses did not fit the data precisely. Consternation was avoided because in the meantime Newton had shown that it was not the ellipses which were fundamental, but the equation of universal gravitation, with force varying inversely as the square of the distance separating two bodies and proportional to the product of their masses.

Universal gravitation *predicted* ellipses, provided that only two bodies were present. If three or more bodies all affected each other, the motion would be more complicated.[1] However, since the Sun is so much more massive than any of the planets, each one moved *almost* as though only the Sun were also present (Fig. 4.2). The other planets caused only small *perturbations* in each other's motions.

These perturbations were understood so well in the eighteenth century that observations of Uranus (discovered by Herschel in 1781) led to the prediction by Leverrier and independently by Adams of another planet, Neptune, which was found in 1846 by Galle in Berlin.

The simulation in this chapter applies the equation of universal gravitation (Eq. 4.1) to each body at each time step using mathematical techniques described in section 4.2.2.

4.2 Theoretical Concepts

4.2.1 Universal Gravitation

Newton proposed the equation

$$F = -GMm/r^2, \tag{4.1}$$

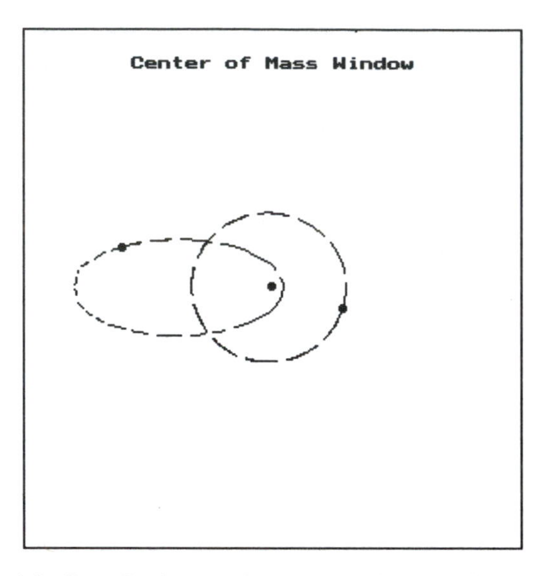

Figure 4.2: Sun, Jupiter, and a comet: almost elliptical orbits.

where M and m are the masses of the two bodies, r is the distance between them, and G is a constant whose value is determined by measuring the tiny gravitational forces between two objects in the laboratory. It is still the primary tool for planetary motion calculations for all bodies of masses comparable to the Sun's mass, only tiny corrections having been introduced in the twentieth century.

This equation is exactly soluble for two-body motion[2,3]; for three or more bodies, approximation techniques must be used. In the last three centuries many such techniques have been developed.[1] These have been further refined since the advent of computers. The projection of the motion of the solar systems hundreds of millions of years into the past, as has been done, requires the utmost refinement of both sophisticated analysis and computing ability.

For the present simulation, we are using a straightforward numerical integration technique. Numerical integration is always an approximation, since an integral is defined in the limit as the number of steps becomes infinite and the step size approaches zero. Any numerical calculation must necessarily use a finite number of steps of finite size.

4.2.2 Approximation Methods

The basic idea of numerical integration is to calculate the forces and accelerations at an initial time, project the velocity and positions of the objects into the future, and repeat the process over and over again. The most elementary set of numerical integration equations for the motion of a single body in one dimension under the action of the force $F(x)$ is

$$a_i = F(x_i) \tag{4.2}$$

$$v_{i+1} = v_i + a_i \Delta t \tag{4.3}$$

$$x_{i+1} = x_i + v_i \Delta t, \tag{4.4}$$

where i numbers the time steps and $\Delta t = t_{i+1} - t_i$. *This quickly runs into trouble, because the velocity and acceleration don't stay constant during the interval* Δt. The simplest way to do a little better is to figure out what the velocity is at the midpoint of the time interval and use that. This is an improvement, but it is not good enough for our purposes.

The technique used in this simulation is called the Runge-Kutta method.[4] It evaluates the accelerations and velocities at several points within the interval to calculate a more accurate approximations. The theory gives us a set of weights or coefficients by which to multiply the results. Fourth-order Runge-Kutta requires the evaluation of four accelerations and velocities at each step; higher-order methods require enough more that they are less efficient.

There is still one difficulty, which is that when orbiting bodies come very close to each other, their speeds are so high that the accuracy of the calculation decreases unless the time step is decreased. This is evident, for example, in calculating the motion of a comet with a 100-year period. When it is far from the Sun, it doesn't move very far in a month, so a time interval of a month is fine for calculation and a shorter time interval would slow the simulation by making unnecessary calculations. However, when the comet comes close to the sun, it may travel 180° around the Sun in a month or less, and one would hardly expect to get an accurate result using such a time interval.

The solution is to use a so-called adaptive integration scheme (Runge-Kutta-Fehlberg),[5] which estimates the error in the calculation and adjusts the time step accordingly. Then a short time step can be used when the comet is close to the Sun and a long one the rest of the time. The size of the error is estimated by comparing the result of a fourth-order calculation with that of a more accurate fifth-order calculation.

Because of the changing step size, bodies in the simulation appear to slow down when they are close together, at the very time when real bodies are traveling at their fastest! A realistic display would be possible but would mean that you would spend a great deal of time watching nothing much happen. You can do that if you want to, because the program gives you the opportunity to change the maximum size of the time step. A preferable solution to the problem of making the real comparative speeds of the bodies evident is to save all the positions of the bodies in a time history, and replay them on demand. That option is provided.

4.2.3 Orbital Elements and Initial Conditions

A considerable amount of rather arcane nomenclature for the parameters of an orbit has developed over the years. Everything about an orbit has a name. An orbit has a major axis, a minor axis, and an eccentricity. If the central body is the sun, the orbit has a perihelion (closest point to the Sun) and an aphelion (farthest point). You can easily deduce the meanings of perigee, apastron, perilune, and the like. The generalized term for the extremes of the orbit is *apse*, as in periapse. The simulation described in this chapter assumes that all the orbits being considered are in the same plane. This is a reasonable approximation to many interesting cases, including the Sun's planets, and greatly simplifies display of the results, although even in the solar system, some moons have orbits greatly inclined to the rest. When creating a new system or adding bodies to a system, you may choose to specify the initial conditions either in polar (r, θ) coordinates or by specifying orbital elements 5–9 as described below.

The complete specification of an orbit in three-dimensional space requires the following[7]:

1. **The Ecliptic**: A standard plane, defined for the solar system as the average plane of the Earth's orbit.

2. **A Standard Direction**: For the solar system, toward the first point of Aries, where the ecliptic enters the constellation Aries.

3. **Longitude of the Ascending Node**: The angle between element 2 and the point where the orbit intersects element 1 going upward (toward north).

4. **Inclination**: The angle between the plane of the orbit and element 1.

5. **Longitude of Periapse**: Angle relative to element 2.

6. **Semimajor Axis**: Half the largest dimension of the ellipse.

7. **Eccentricity**: The displacement of the center of the ellipse from the focus occupied by the central gravitating body, divided by the semimajor axis. It ranges from 0 for a circle to 1 for a parabola and larger for a hyperbola, which is an orbit that escapes from the system.

8. **Time of Periapse Passage**: Determines the position in the orbit at any time once the period is known. The program settings use the equivalent: position angle relative to periapse at time zero.

9. **Masses of the Bodies**: These along with element 6 determine the period.

Since the simulation described here is two-dimensional, only elements 5–9 need be specified for our purposes, with element 5 given relative to the x-axis as the standard direction. These determine uniquely the initial r, θ, and velocity, although you must also specify whether the motion is clockwise or counterclockwise (astronomers would say direct or retrograde).

No body in a many-body system under gravitational forces actually pursues an elliptical orbit. At any given time, however, the motion may be approximated by an ellipse, and the ellipse which best fits the motion at any given time is called the *osculating ellipse*. This is the ellipse which a body of this velocity and position would follow if all the other bodies except one were removed. Obviously, this is meaningful only if one body in the system has most of the mass, so that the effects of the other bodies are perturbations. These perturbations can then be viewed as changing the parameters of the osculating ellipse (Fig. 4.3).

The simulation described here displays the numerical values of the periapse and eccentricity of the osculating ellipse at time zero. It does not attempt to keep the numerical values up to date (Exercise 8). They are shown only when there is enough room on the screen: one or two windows displayed.

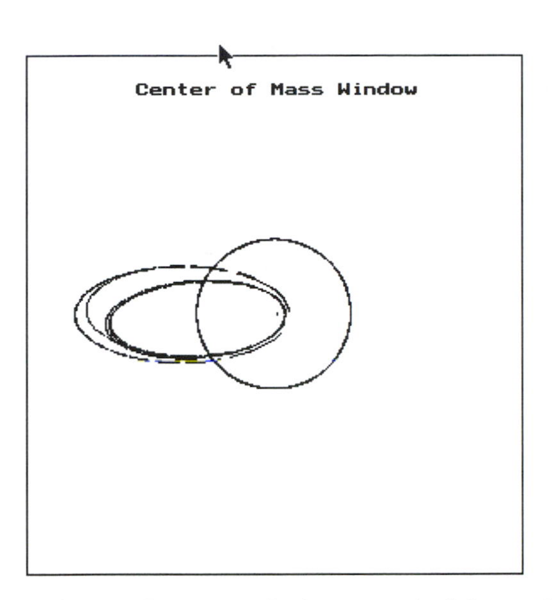

Figure 4.3: Sun, Jupiter, and a comet: Jupiter perturbed the motion of the comet when they came close to each other.

4.2.4 Chaos

The observed stability of the solar system over billions of years is remarkable in view of the fact that even three-body motion under the influence of gravity can result in wild orbits, such as those of the comet in the **Binary Star & Comet** system (Fig. 4.4). The solar system orbits are tame because the masses of all the bodies except one are small. In the current decade it has been established that even the solar system is chaotic (chaos is discussed in chapter 3). This was first established for Pluto,[8] and very recently for the solar system as a whole.[9] This does not mean that planets will go flying off wildly, only that their positions in orbit, and the elements of the orbit, will be increasingly different at increasing times for nearly identical initial conditions. In the case of Mars, the inclination of its axis of rotation seems to have varied widely, with a resulting drastic effect upon seasons.

4.2.5 Phase Space and Poincaré Diagrams

There is a full discussion of these concepts in chapter 3. The simulation described in the current chapter does not present any phase space orbits as such. The Poincaré diagrams presented here are those of a third body in the presence of two more massive bodies, and are in the rotating system in which the two massive bodies are stationary. The position of the body in x, v_x space is plotted whenever y has some fixed value, usually zero.

4.2.6 Computational Approach

The computational technique is the same whatever the system being modeled. Equation 4.1 is applied over and over again to however many bodies there are. The

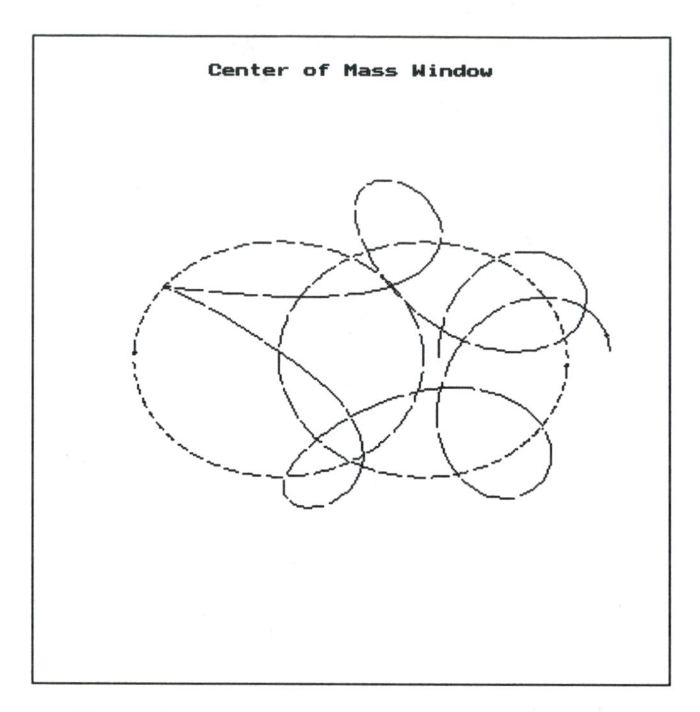

Figure 4.4: Two stars of equal mass and a comet.

resulting acceleration is integrated using an adaptive Runge-Kutta-Fehlberg routine discussed more fully in section 4.2.2 which adjusts the time step to maintain accuracy when bodies are close together. The result is that whenever bodies approach each other, their motion appears to slow down, just the reverse of what an observer viewing the real system sees. To see the motion in real time, the user must choose **Replay**, which retraces the orbits using a stored position history. The motions are restricted to a plane, which is a reasonable approximation to many systems of interest. It is possible to remove this restriction upon the calculation (Exercise 10), but it is far more difficult to display the results if the motions are far enough from a plane that projection is not satisfactory.

The simulation allows an unlimited number of bodies, at most five of them with a mass greater than zero. That means that the rest do not perturb the motion of any other bodies. This allows a realistic approximation to a large class of problems where the mass of most of the bodies (planets, comets, satellites, etc.) is negligible. It clearly excludes globular star clusters, but those are far beyond the computational capabilities of the computer systems on which the simulation is expected to run.

The several models available from the menu are really just sets of initial conditions for the simulation. Facilities for entering entirely different initial conditions are provided, using a combination of data entered from the keyboard with positions and velocities selected by the mouse and refined from the keyboard if necessary.

Thrust can be applied to one body in each system, the highest numbered one, using the arrow keys. This feature is disabled by default for most systems, and can be enabled from the menu.

4.3 Types of Orbital Systems

In any of the following systems, the user can display other windows or change their scales at any time, and can "replicate" a body, starting a pair or a swarm of bodies from nearby positions in phase space to see how their subsequent orbits relate to each other. Divergence of orbits is a sign of chaos.

4.3.1 Sun-Earth-Moon

The default display of this system views it through four windows: The **Center of Mass** window shows the entire system. The Earth and Moon are indistinguishable from each other on this scale. The **Earth** window and the **Rotating** window both have scales chosen to display the Moon suitably. Notice that the Earth moves in the **Rotating** window (Fig. 4.5). The Earth changes its distance from the Sun, while the center of the window is at a fixed distance from the Sun.

The **Sun** window is basically the same as the **Center of Mass** window except that it is magnified sufficiently to distinguish the two orbits. That means that only a portion of the orbits is visible, and the initial conditions are such that you have to wait for the bodies to go by. The location is chosen to fit as large a piece of the orbits as possible into the diagonal. The choice of scale is dictated by the desire to display the difference between the Moon's orbit and the orbits of Jupiter's satellites (Exercise 2), and is a compromise between showing enough of the orbits, and keeping them separate.

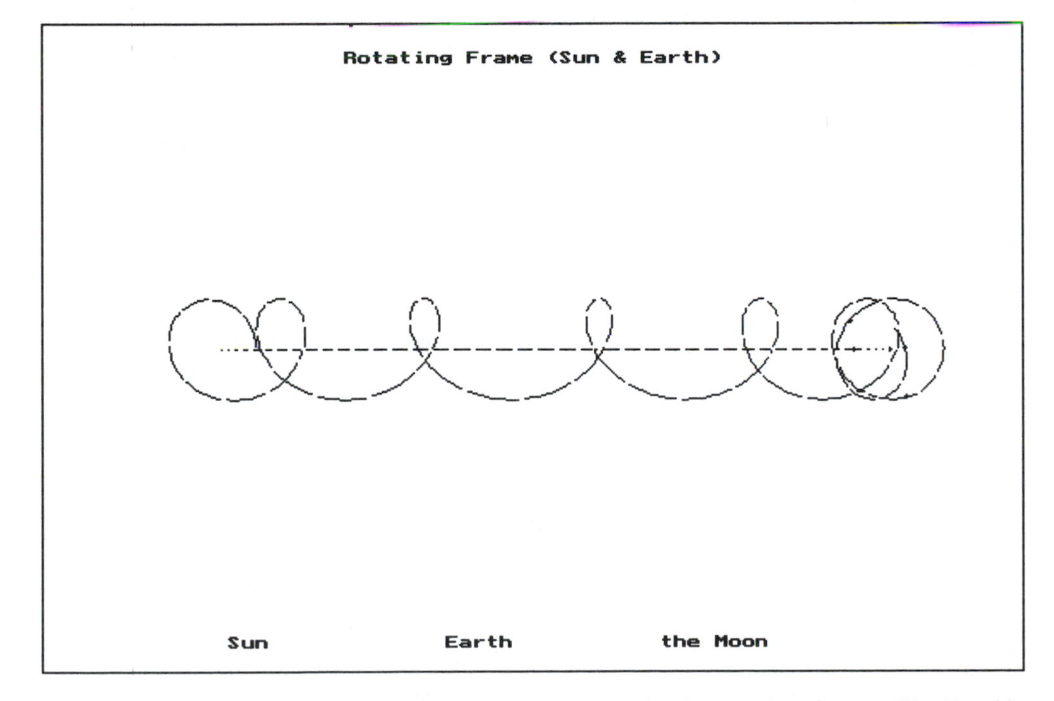

Figure 4.5: Earth (short dashes) and its moon in the rotating frame. The Earth's distance from the Sun changes and carries the moon along.

4.3.2 Solar System

Two windows are provided initially: one showing the **Inner Planets**, and one showing the entire system, but labeled **Outer Planets** window since the inner planets are too close to the Sun to show much detail in this view.

The initial positions are chosen randomly, rather than according to any table of planetary positions. The eccentricities of the orbits are correct, but the major axis orientations are random. Exercise 4 asks you to substitute realistic initial conditions at a time chosen by you.

4.3.3 Shuttle Docking

The system is composed of the Earth, a space station, and a space shuttle. Thrust can be applied to the shuttle in either the x or y directions (in the rotating system) using the **arrow** keys. The challenge is to approach the space station and match velocities; it is a major challenge. The method used by the astronauts seems to have been to always circularize the shuttle orbit after making any change.

The default windows are the **Center of Mass** window (identical to the **Earth** window in this case) and the **Rotating** window magnified to an appropriate scale. The station and the shuttle are initially about 50 km apart.

4.3.4 Sun-Jupiter-Comet

This is provided to demonstrate that a near approach to Jupiter makes major changes in the orbit of a comet. Many real comets have aphelia near the orbit of Jupiter as a result of many such encounters.

The default windows are **Center of Mass** and **Rotating**. The default size of the body dots is larger than in the other simulations in order to emphasize the body positions for this system.

4.3.5 Sun, Jupiter, and Moons

This system is analogous to the Sun, Earth, and Moon system, and starts up with the same four windows. The four Galilean moons are shown. The magnified orbits of the moons are very different from that of the Earth's moon. Exercise 2 investigates the reason.

4.3.6 Lagrangian Points

Lagrangian points are points of equilibrium in an orbital system.[10] The system simulated is composed of the Sun, Jupiter, and three small bodies in orbits similar to Jupiter's, one at 60° away from Jupiter, one at 90°, and one at 180°. One of the small bodies is at a point of stable (dynamic) equilibrium,[6] and one is at a point of unstable equilibrium. The latter has a fascinating orbit if you follow it long enough; it eventually makes a close approach to Jupiter.

The default windows are the Jupiter window and the Rotating window. All the interesting action is in the latter; the former is included just to make it immediately clear to the observer that the objects are rotating.

4.3.7 Binary Star and Comet

Two stars and a third body whose exact nature doesn't matter. The initial windows are **Center of Mass**, **Star A**, **Star B**, and **Rotating**. The comet remains essentially bound to one star, which is more evident in some windows than in others. I chose to give the center of mass a non-zero velocity, so that there are motions worth viewing in the **Universe** window.

Start the comet at different positions and velocities (**Move Body** on the **Choices** menu). Also use **Replicate** to see what happens to bodies starting from slightly different places.

4.3.8 Retrograde Motion

This system is composed of the Sun, Earth, and Jupiter, shown initially in the **Center of Mass** and **Earth** windows. The line of sight from the Earth to Jupiter is drawn in both windows, graphically illustrating the difference between the two views and the reason why Jupiter appears to move backwards at times when seen from the Earth. The line of sight leaves ghosts behind it periodically to further assist the visualization.

4.4 Using the Program

4.4.1 Setting the System in Motion

All systems are stationary when first selected. Set them in motion by pressing the F2 key; stop them again by pressing the F3 key or clicking the mouse button. The motion can be replayed from the menu. It is automatically replayed after any changes of windows. This replay uses positions saved in memory, and so shows realistic speeding up and slowing down.

4.4.2 Graphical Display

Each system may be viewed from any of several reference frames, either one at a time, or up to six simultaneously. The available frames include those in which the center of mass of the entire system is stationary, as well as those in which any of the first three bodies is stationary. This can dictate the order in which bodies are numbered. The motion can also be viewed from a frame stationary in the universe, and from the so-called rotating frame described in the next paragraph. The final view available is that of the Poincaré diagram. These views are referred to in the program as "windows" on the motion.

The rotating frame is one in which the radius vector between body 1 and body 2 remains constant in direction. This is particularly useful in viewing the motion of a planet and its satellites (Fig. 4.5), or in studying the motion of objects near the Lagrangian points (defined below) of a system. Multiple rotating-frame windows with different centers are allowed, making it possible to zoom in on two different parts of the system and compare their motions on a magnified scale.

Facilities are provided for zooming into or out of any window, thus changing its scale. The default change of scale factor is ten, but a facility to change this easily is provided. Zooming out centers the new window on the old one. Zooming in requests the user to pick the center of the new window using the mouse.

4.4.3 Star Trails

There are a number of choices surrounding the display of an orbit and the current position of a body. Since no single choice is best for all situations, the user has control over them. The default choice is that the past orbit is shown, and so is the current position of the body—as a small dot every ten time steps. When this dot is erased, it also erases a portion of the past orbit. In many cases, this leaves nicely spaced gaps outlining the time history of the motion.

If the body is moving slowly, the motion in ten time steps is so small that the entire track is erased. This happens to some bodies in the **Lagrangian Points** system, for example. Redrawing from the history did not seem like a good solution, since the number of steps redrawn would have to depend on the speed. To see the track of a slowly moving body, **Replay** the motion. **Replay** suffers from the opposite problem—that the gaps disappear for slow motions, since they are now a constant length in time instead of a constant length in space. Displaying the dots in **Replay** would not be a good solution, since they then fall on top of each other for slow motions.

You have the option under **Settings** on the **Choices** menu of changing the size of the dots or turning them off, or turning off the past track.

4.4.4 Changing Parameter Values

- **Settings** on the **Choices** menu gives you the ability to change several parameters related to the graphical display and to the overall integration, including the integration precision.

- **Change Parameters** allows you to change the name, mass, position, and velocity of any body at the current time. If you wish to change the initial position, reset the current time to zero if necessary by using **Restart**.

4.4.5 Program Menus

- **Files**: Allows access to files that contain initial conditions and system parameters, allowing you to save initial conditions as well as entire new systems with new masses, names, and orbital parameters.

  - **About Orbiter**: Brief description.
  - **About CUPS**: Brief description of the project.

- **Configure**: Allows you to change the path to the directory in which system parameter files are stored (shortcut key **Alt-d**), change the colors (shortcut key **Alt-c**), or check the amount of memory available to the program (shortcut key **Alt-m**). You can change the colors to black-and-white to print the screen on a black-and-white printer.
- **Read System From File**: Reads a system with initial conditions from a file.
- **Save System to File**: Saves the current state of the system you have selected to a file, so that the present conditions become the initial conditions in the file. Whatever changes you have made, including adding or replicating bodies, will be saved.
- **Exit Program**: Can also be accomplished by pressing the shortcut key **Alt-x**.

• **Choices**: This menu collects all the choices that can be made about a system and about the display, other than picking an entirely new system or changing the windows through which it is viewed.

- **Clear & Continue**: Clears all windows; the motion then continues from where it was when the **Clear** request was made.
- **Restart**: Resets the initial conditions to what they were when the system was loaded or the last body was added. If a body has been replicated or moved, you will be asked whether you want the original system or the modified one.
- **Replay**: Erases the screen and plots the motions from the replay history saved in memory. The replay history is a circular buffer, and its size depends upon the amount of memory available. If your motion has continued for a long time, or the memory available is small, the replay may not extend back to the beginning.

  The replay proceeds at a constant ratio to real time, since the data are saved at equal time intervals even when the computation is using smaller steps for accuracy. The replay track will have gaps in it at equal time intervals if the **Settings** option **Show Body Positions** is true. These gaps will tend to disappear when the motion is slow, instead of the tracks disappearing at those points during the original computation. Replay will show tracks even if the option **Show Past Track of Bodies** is false.
- **Replicate Body**: Allows you to make up to eight new replicas of a body distributed in phase space near the body you choose. You will be asked to choose which body, how many replicas, and how they are distributed in phase space.
- **Settings**: Allows you to set a number of parameters affecting the computation and the display. The choices are—

  * **Time Step**: The time interval at which the graphic and numerical displays are updated. The integration may use a smaller step for accuracy, but this does not affect the display.
  * **Largest Time Step**: The integration routine will not increase its time step beyond this. It has no effect if it is larger than the **Time Step**, and I have been unable to think of circumstances in which it is useful to make it smaller.

* **Automatic Replay After Changing Number of Plots**: When you change the number of windows, replay will occur automatically if this option is true. Turn it off if you have a long history that you don't want to wait to see, especially if you are trying out some window choices to see what they are like.

* **Automatic Replay When Choosing Next Plots**: One of the window choices allows you to flip quickly from one window to the next in a predetermined order, with automatic replay. Turn it off if it is annoying.

* **Size of Dots Marking Present Body Positions**: Allows you to make the dots larger if you have trouble seeing them, or smaller if they are obtrusive. Unfortunately, the range of sizes provided by Pascal is not as fine as one would wish. Size 1, which is the default for most of the systems, is usually a little too small, and size 2, which is the default for the Sun-Jupiter-Comet system, is rather too big.

* **Show Body Positions**: If you turn this off, you will see only the tracks of the bodies without any dot to mark the "current" position, and with no gaps in them marking past time intervals. The latter will be true of the replay as well, which shows no dots for either setting of this choice. The dots are only plotted every ten time steps, both to save time, and to keep the gaps from obscuring the tracks.

* **Show Past Track of Bodies**: If you turn this off, only the current positions of the bodies will be shown, unless you turn that off as well. Turning both off speeds up the computation by about 20%, but you would need to use **Replay** to see what happened. **Replay** always shows the tracks, regardless of the setting of this parameter.

* **Display orbital elements of body *x* about body *y***: You can change *x* and *y* to determine which orbital elements are displayed.

* **Absolute Integration Error**: Sets an absolute upper limit to the allowable integration error. The integration routine adjusts the time step it uses (not the one displayed) to keep the error of every variable under this limit. The absolute error limit controls when the values of some variables are so close to zero that the relative integration error is a smaller number than the absolute error limit.

* **Relative Integration Error**: Sets a fractional error limit: The error in no variable can be larger than this fraction of the value of the variable. This limit controls when the values of the variables are large.

- **Reverse Time**: This allows a check on the accuracy of the integration by running it backward to see if it ends up at the starting point. You can also set up a configuration of bodies and see what past history would have led to it. In the orbital realm motions are reversible, but the breaking of an egg (or a planet) is not. The relation between the two is still the subject of debate and publication.

- **Change Parameters**: Allows you to change the position and velocity of any body, and the mass of any of the first five bodies. You have a choice between polar coordinates and orbital elements (section 4.2.3). When you select this menu item, a submenu appears listing the bodies in the system, as well as the system itself. The last allows you to change the value of G and the time step. You can change as many bodies as you like, selecting **Done** when you are through.

- **Move Body (Mouse)**: Allows you to point to any body with the mouse and move it wherever you like. While you are moving it, its distance from the nearest other massive body appears at the bottom of the screen. When you release the mouse button, a line appears indicating circular velocity relative to the nearest other massive body. This is calculated as if no other bodies were present. While that is unrealistic if other bodies are nearby, it does provide a reference scale for reasonable velocity magnitudes.

  You can then accept the "circular" velocity with the F3 key or reverse its direction with the F4 key. You can also move the tip of the velocity line with the mouse to whatever direction and magnitude you like. The ratio to the reference "circular" velocity appears at the bottom of the screen.

- **Add Body**: After choosing this item, click the mouse where you want the new body to appear; if you hold the mouse button down, you will see how far the body is from the nearest massive body and you can adjust its position. You then proceed just as with **Move Body**, except that at the end an input screen appears, allowing you to name the new body and make fine adjustments in its parameters.

- **Allow Thrust**: Activates the thrust controls for the highest-numbered body in the system. Some systems do not allow this option; if you create a new system it is always allowed. Once thrust is allowed, the **arrow** keys each apply one unit of thrust in the direction of the arrow. The size of the unit can be changed by the **Page Up** and **Page Down** keys.

- **Plots & Zoom**

  - **Choose One Window**: Displays a list of the seven possible windows to choose to display full screen.

  - **Next One Window**: Rotates through the possible windows in a set order, beginning with the Center of Mass window.

  - **Choose Two (Four, Six) Windows**: Brings up the same set of choices as **Choose One Window** to choose the specified number. It is possible to choose the **Rotating Window** more than once so that you can zoom in on different bodies in different windows, for example, Jupiter plus any of the small bodies in the **Lagrangian Points** system.

  - **Mark Center of Mass**: Places (or removes) a cross at the center of mass of the system. The cross does not appear until you either choose **Restart** or set the system in motion.

  - **Full (Part) Screen**: Widens a single window to fill the whole screen, covering the numerical parameter display. **Part Screen** will then restore the original display. If **Full Screen** is chosen while multiple windows are open, you will be asked which one you wish.

- **Zoom In**: Allows you to magnify a selected window. A list of open windows appears. Select one, and click on the point you wish for the center of the magnified display. You can repeat this for the same window to get more magnification, or for a different window. When done, click **Done**. You can change the magnification factor by typing in the input panel at the lower right of the screen.

- **Zoom Out**: Reverses **Zoom In**, except that the center of the old window is taken as the center of the new one. The current magnification factor is used, unless you change it.

- **Default Scales**: Resets the scales and centers of all windows to the values which they had when the system was first selected, without erasing the orbit history or changing the choice of windows.

- **Systems**
  The selections on this menu allow you to choose one of the systems listed and described in section 4.3. In each case, the windows displayed initially are chosen to be appropriate for the particular system, but may be changed at any time.

  - **Various Systems**: See the listing in section 4.3.

  - **Create New System**: This allows you to create any system you choose. The first screen that appears allows you to set the scale of the plot at the bottom, as well as naming the first body and choosing its mass, position, and velocity relative to the center of mass of the system.

    When you have accepted the parameters of the first body, you are asked to name and set the mass of the second body. When you do so, a window will appear allowing you to place the body and its velocity with the mouse, just as in **Add Body**, with the same possibility of making fine corrections to the numbers when you have done so.

    After you specify each body, the screen clears, and you are offered the choice (using F-keys) of adding another body or finishing.

- **HELP!!**: Changes the menu into a **Help** menu. Selection of any menu item then displays one or more screens of information about that item, and this can be done repeatedly. Leave the help system by clicking the mouse anywhere but in a menu, pressing **Enter**, or selecting **Quit Help**.

4.5 Exercises

If you have facilities to print a graphics screen you may print out some orbits. Whether you do or not depends on software provided locally. If you are running under Windows, **Alt-PrintScreen** will copy your entire screen to the clipboard. You can then paste it into Windows Paintbrush (after **Zooming Out**) and print it from there. Change the colors to black-and-white before you do any of this unless you have a color printer.

4.1 **Origin of Kinks**

(a) Set the **Lagrangian Points** system into motion and observe the wiggles in the orbits of the small bodies in the **Rotating Frame** window.

(b) Write down a possible reason for the existence of these wiggles.

(c) What parameters might you change to test the correctness of your hypothesized reason? Write down the changes and what you expect to result from them.

(d) Change the parameters. Do they have the effect you expected?

(e) If the effect was not what you expected, repeat b–d until you are satisfied that you cannot do any better. Hand in records of all steps; the process is more important than whether or not you find a correct reason.

4.2 **Different Moon Orbits**

(a) Set the **Sun, Jupiter & Moons** system in motion and observe carefully the orbit tracks in the Sun window. Make a sketch of them. Do the same for the **Sun-Earth-Moon** system. Describe the difference(s) between the orbit tracks in the Sun window. Which of Jupiter's moons has an orbit most like that of the Earth's moon?

(b) Write down a possible reason for the difference. What could you change that would make the orbit tracks more similar?

(c) Make the appropriate change(s). Does it have the expected result?

(d) If the effect was not what you expected, repeat b and c until you are satisfied that you cannot do any better. Hand in records of all steps; the process is more important than whether or not you find a correct reason.

4.3 **Effect of Changes in Initial Position**
The **Lagrangian Points** system is a good one to use for this, and so is **Binary Star and Comet**, but you may wish to try others.

(a) The **Replicate Body** facility offers four different ways of changing the initial position of a body: changing x, y, v_x, or v_y. Consider a body on the x-axis; predict what the effect of each of these will be and how they will differ from each other, based on your knowledge of orbital mechanics. Write down your predictions to hand in; do not change this record after you see what happens.

(b) Try out each of the four cases using the simulation. Record what happened and how it agreed or disagreed with your prediction.

(c) Revise your predictions and repeat b.

4.4 **Realistic Solar System**

(a) Consult tables of planetary orbits to find out the correct positions of the planets in their orbits and the positions of their perihelia at some time of your choice. Use **Change Parameters** on the **Choices** menu to set them into the simulation. Use **Save System to File** before you run it.

(b) Calculate the x-y positions at some time of your choice of at least two of the planets, and compare them with what you see on the screen. Hand in the calculated positions along with the saved system file on a disk.

4.5 **Search for Chaos**

Use the **Replicate Body** facility to look for evidence of chaotic motion. You are most likely to find this with three bodies of comparable masses, or in the **Binary Star and Comet** system. You can create the former by changing the mass of the comet in the latter. Try different starting points for the third body. What you are looking for is orbits of bodies that start out close together in phase space and get further and further apart as time goes on. Record and hand in the initial positions you try along with a statement of what happened for each.

To make the time spent worthwhile, even if you do not find evidence for chaos, describe the way in which the resulting orbits differ for different starting places of the bodies.

4.6 **Poincaré Maps**

(a) Select the Sun, Jupiter, and Comet System. Select two windows: Rotating and Poincaré. Set the system into motion, and each time a point appears in the Poincaré window, stop the system and satisfy yourself that you understand why the point appeared where it did. Hand in a written explanation with relevant sketches for at least two points.

(b) Let this system run for a while and print a copy of the screen, or at least make a sketch.

(c) Create a Poincaré map of the Sun, Earth, and Moon system in a similar fashion.

(d) Create a Poincaré map of the Binary Star and Comet system in a similar fashion.

(e) Which (if any) of these systems is clearly in a periodic orbit? Give reasons for your answer.

(f) Which of the systems (if any) might be in chaotic motion? Give reasons for your answer. Do you have enough evidence to be certain? Why or why not?

4.7 **Orbit Pyrotechnics**

(a) Move the comet in the **Binary Star & Comet** system to form an approximate equilateral triangle with the two stars, and give it a velocity toward the center of the triangle, about 1.0 times circular velocity as noted at the bottom of the screen. Start the motion and make a sketch of the orbits.

(b) Use **Change Parameters** to change the mass of the comet to 1.0 M_Sun. **Restart** the system using the **System Modified With Mouse** option, and again sketch the orbits. Describe the difference. Why have the orbits of the two stars changed?

(c) Repeat b for several masses between 1.0 and 0.1 M_Sun.

(d) What is the effect of now changing the direction of the initial velocity of the comet? Try as many different directions as you like. Give a reason for choosing whatever mass you did for this part.

4.8 **Change Replication Distribution in Phase Space** (Programming Project)
When a body is "replicated," the new bodies are placed at offsets in a Cartesian coordinate system. That is, the offsets are in the x direction, y direction, in v_x, or in v_y. Change this to a polar coordinate system, so that the offsets are in r, θ, v_r, or v_θ.

This requires modifying one line of the Procedure **ManyProbesInit** in the file ORBHELP.PAS. Use the editor to search for the following comment: "{Adjust the values for the new bodies}". The line to be modified is the second line following, starting with "Y.Put." The positions and velocities of the bodies are stored in the vector Y in the order x_1, y_1, ... x_n, y_n, v_{x_1}, v_{y_1}, ... Function Y.Value(n) returns the value of the n-th element of Y, and Y.Put(n, V) puts the value V into the n-th element. Be careful not to change ib, ispread, and Delta. Much of the logic of this procedure is setting their values properly for the various cases.

4.9 **Osculating Elements** (Programming Project)
Modify the display of the orbital elements so that it periodically brings them up to date as the current osculating elements. The data are displayed by Procedure **ShowData** in the file ORBPHY.PAS. The quantities that need to be recalculated are the eccentricity, perihelion, aphelion, and angular momentum. The necessary data are found in the vector Y, whose structure is described in Exercise 7.

4.10 **Change Force Law** (Programming Project)
Change the force law in Procedure **N_Body_Force** in file ORBPHY.PAS to a different power of r, or a sum of powers. Describe the resulting changes in the orbit of some particular body or bodies.

4.11 **Inclination to the Ecliptic** (Major Programming Project)
The simulation as distributed assumes that all the bodies move in a plane. Modify the force equations and variables to remove this restriction, but display the projection of the motion on the x-y plane. Does the motion of a body that has a considerable inclination to the x-y plane differ much from that of a body with the same initial x and y positions and velocities but zero z position and velocity?

The main changes need to be made in Procedure **N_Body_Force**, which is in the file ORBPHY.PAS (as are the other procedures you need to change). This procedure calculates the values of the elements of YP, the vector of the derivatives of Y, both required by the numerical integration routine Procedure **StepRKF** (in CUPSPROC.PAS). The structure of the vector Y is described in Exercise 7. The third z-component will behave just like the other two. You will need to modify Procedures **ThreeBodyInit** and **GeneralInit** to give initial values to the z position and velocity (look at how they do it for x and y). You must also change Y.Resize, and YP.Resize in the latter two procedures to make the vectors half again as large to include the z values. You might also modify the Procedure **SetBodyParameters** to allow changing them while the program is running, but that requires understanding data input screens and you may find it preferable just to change the initialization routines and recompile. In order to make **Restart**, **Add Body**, and other options work properly, you would have to do considerably more programming; it is suggested that you keep the project manageable and just not use these options.

References

1. Blanco, V. M., McCuskey, S.W. *Basic Physics of the Solar System*. Reading, MA: Addison-Wesley, p. 169ff, 1961.

2. Baierlein, R. *Newtonian Dynamics*. New York: Mc-Graw Hill, 1983.

3. Fowles, G. R. *Analytical Mechanics*. New York: Saunders College Publishing, 1986.

4. Press, W. H., et al. *Numerical Recipes*. Cambridge: Cambridge University Press, 1986, p. 550ff.

5. Maron, M. J. *Numerical Analysis*. New York: Macmillan, p. 344ff, 1982.

6. Baierlein, *op. cit.* p. 181ff.

7. Taff, L. G. *Celestial Mechanics: A Computational Guide for the Practitioner*. New York: Wiley, p. 30ff, 1985.

8. Sussman, G. J., Wisdom, J. The numerical evidence that the orbit of Pluto is chaotic. Science **241:**433, 1992.

9. Sussman, G. J., Wisdom, J. Chaotic evolution of the solar system. Science **257:**56, 1992.

10. Blanco, McCuskey, *op. cit*, p. 179ff.

5

Rotational Dynamics of Rigid Bodies

Randall S. Jones

5.1 Introduction

To obtain a quick introduction to this simulation see the ROTATE walk-through in the Appendix A.

The rotation of rigid objects is an important subject that may be the most conceptually difficult topic in classical mechanics. There are several reasons for this. First, the motions of wobbling plates and precessing tops are not intuitively obvious to most of us. Furthermore, discussions of rotations usually require visualizing quantities in three dimensions. Angular velocity, angular momentum, and torque by their definitions require working in directions that lie out of the plane of the motion. Finally, visualization is made more complicated by the fact that the equations of motion (Euler's equations) are most easily represented in a reference frame tied to the rotating body. Translating the resulting derived quantities back to the fixed lab frame is often not trivial.

This simulation is designed to aid in the visualization of rotational motion. It will allow you to observe the three-dimensional (3-D) motion of rotating objects in a controlled fashion, running the simulation faster, slower, or in reverse, while displaying the corresponding evolution of the angular velocity, the angular momentum, and the torque. It will display the motion from the fixed frame and from the body frame to help in understanding the translation between these two descriptions of the motion. By using the stereographic feature of the program, you can create a genuine 3-D image of the motion of these quantities.

It is important to remember that in *classical* mechanics, the equations of motion for rotational motion provide no new physics. The motion of any rotating object can always be understood strictly from a consideration of the linear form of Newton's second law ($F = ma$) applied to the individual objects making up the rotating body. In some sense, then, the rotational quantities (angular momenta and torques) and their differential equations are merely convenient ways of simplifying the analysis of these systems. At the same time, of course, it is clear that angular momentum has some fundamental significance, as indicated by the fact that it is a conserved quantity when no net torques are applied. A number of the problems at the end of this chapter are designed to help develop an intuitive understanding of rotational motion, both in terms of the rotational quantities and in terms of the linear equations applied to individual point masses.

5.2 Theoretical Concepts

Rotational motion and the equations governing their time development are discussed in detail in all of the standard mechanics texts, so derivations are not reproduced here. Rather, I will attempt to give an intuitive account of the quantities used to represent rotational motion and their equations. The notation used here is that of Marion and Thornton,[1] but it does not differ significantly from that of other authors.

A rigid body is usually treated as a collection of point masses, m_i, located at positions, $\vec{r}_i$, measured relative to some coordinate system attached to the body. An important fact in describing rotational motion is that for any rigid body there always exist three mutually perpendicular axes (the *principal* axes), intersecting at the center of mass, around which the body will rotate uniformly. That is, if the body is given a rotational velocity about one of these axes, and is subsequently free from external torques, the rotation will continue about that axis and the orientation of the axis will remain fixed in space.* The existence of principal axes is not surprising for highly symmetric objects such as a football or Frisbee, but in fact, these axes exist even for complicated objects such as the Hubble space telescope.

Rotation about an axis that does not correspond to a principal axis will never be simple. If the body is constrained to rotate about that axis by some mechanical axle, then a torque must be applied to the body by the axle to maintain the orientation of the rotation axis. Alternatively, if no torque is provided, the rotation axis will change orientation, often in a complicated way.

The principal axes are so important that a rotating object is almost always described relative to a coordinate system formed by these axes. This coordinate system is usually referred to as the *body frame* or *rotating frame*. Since this coordinate system is attached to the object, it must move with the rotating object and is therefore a non-inertial reference frame. Throughout this chapter primed symbols (i.e., $\vec{r}'$) will be used to represent vectors in the fixed frame (sometimes called the lab frame), and unprimed symbols (i.e., $\vec{r}$) will be used to represent vectors in the body frame. The origin of the body frame might be translating with a uniform velocity with respect to the lab frame, but since this generates no new physics that case will be ignored.

There are four fundamental quantities associated with rotational motion of rigid bodies: The *angular velocity*, the *angular momentum*, the *torque*, and the

*Note that this rotation may not be stable in the sense that a small perturbation of the rotation away from the principal axis may cause the rotation axis to change its orientation.

moment of inertia. The angular velocity is, in effect, defined by the expression giving the velocities of the particles making up the rigid body:

$$\vec{v}_i' = \vec{\omega} \times \vec{r}_i' , \tag{5.1}$$

where $\vec{v}_i'$ is the velocity of particle i in the fixed frame. The fact that a single angular velocity will suffice for all the particles is actually a definition of rigid body. The angular momentum is defined by

$$\vec{L} = \textstyle\sum_i \vec{r}_i' \times \vec{p}_i' , \tag{5.2}$$

where $\vec{p}_i' = m_i \vec{v}_i'$ is the linear momentum of each particle. The external torque applied to the object is determined by the external forces $\vec{F}_i$ applied to each particle, according to

$$\vec{N} = \textstyle\sum_i \vec{r}_i' \times \vec{F}_i . \tag{5.3}$$

Note: We will often have occasion to speak of $\vec{\omega}$ and $\vec{L}$ *in the body frame*. This does not mean that we are measuring these quantities in the body frame (clearly $\vec{\omega}$ and $\vec{L}$ must be zero in a frame fixed to the body). Rather, it simply means that we are specifying the vector components along the instantaneous directions specified by the body axes.

The final important quantity in describing any rigid body is the moment of inertia. Intuitively, the moment of inertia represents the "resistance" of the object to angular acceleration, just as the mass represents the "resistance" of the object to linear acceleration. In the body frame, the moment of inertia can be specified by three quantities—I_x, I_y, I_z—for angular accelerations about the x-, y- or z-axis, respectively. These quantities are determined by

$$\begin{aligned} I_x &= \textstyle\sum_i m_i(y_i^2 + z_i^2) \\ I_y &= \textstyle\sum_i m_i(z_i^2 + x_i^2) \\ I_z &= \textstyle\sum_i m_i(x_i^2 + y_i^2) . \end{aligned} \tag{5.4}$$

Note that the moment of inertia is not a vector, but is in fact a rank 2 tensor containing nine elements. With the choice of the principal axes as the body coordinates, only the three elements defined above are non-zero. Since we consider only problems with the body axes aligned with the principal axes, we do not need the tensor generalization. Refer to textbook chapters on rotational motion for more details.

If rotation occurs only about a single principal axis, (e.g., the z-axis), the angular momentum is given by the relation

$$L_z = I_z \omega , \tag{5.5}$$

and the equation of motion is simply the standard freshman physics result:

$$N_z = I_z \frac{d\omega}{dt} . \tag{5.6}$$

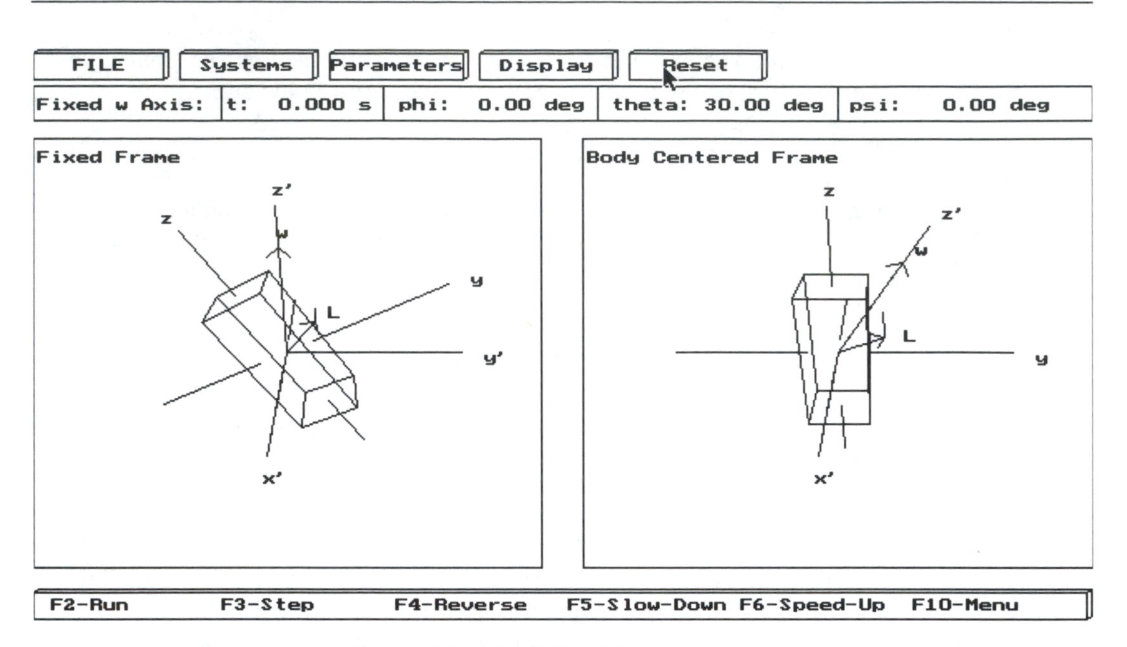

Figure 5.1: The ROTATE user screen.

Note that we do not need to distinguish between the body and fixed frames in this equation since the z- and z'-axes coincide in these simple cases.

If the rotation occurs around an axis that is not a principal axis, the motion will be more complicated and we need a notation to describe this motion. A standard choice for representing the instantaneous orientation of an object is the Euler angles, defined as follows: An initial orientation of the object is chosen, with the body-frame coordinates aligned with the fixed-frame coordinates (see Fig. 5.1). A subsequent orientation of the object (and body frame) is represented by the following three rotations required to rotate the object from its initial orientation into its new position: a rotation of the object about the x'-axis by an angle θ; a rotation of the object about the z'-axis by an angle ϕ; a rotation of the object about the new z-axis by an angle ψ. Note that the angles θ and ϕ are almost the same as the corresponding spherical variables. Because the rotation by θ is about the x-axis, rather than the y-axis, the Euler angle ϕ is always $\pi/2$ greater than the spherical angle ϕ (see Fig. 5.1). A rotation matrix may be established according to

$$M_{Euler} = M_z(\phi) \cdot M_x(\theta) \cdot M_z(\psi) \tag{5.7}$$

$$M_{Euler} = \begin{bmatrix} \cos\psi\cos\phi - \cos\theta\sin\phi\sin\psi & -\sin\psi\cos\phi - \cos\theta\sin\phi\cos\psi & \sin\theta\sin\phi \\ \cos\psi\sin\phi + \cos\theta\cos\phi\sin\psi & -\sin\psi\sin\phi + \cos\theta\cos\phi\cos\psi & -\sin\theta\cos\phi \\ \sin\psi\sin\theta & \cos\psi\sin\theta & \cos\theta \end{bmatrix}, \tag{5.8}$$

which rotates the components of each mass particle constituting the object into its new position. This matrix also transforms the the body-frame components of a vector to the fixed-frame components. The transpose of this matrix will transform fixed-frame components to body-frame.

The rotational velocity of the object can be represented by the time derivatives of the Euler angles $\dot{\phi}$, $\dot{\theta}$ and $\dot{\psi}$, but it is more convenient to represent this velocity in terms of the components of the instantaneous angular velocity vector. The relationship between time derivatives of the Euler angles and the angular velocity is as follows:

$$\begin{aligned} \omega_x &= \dot{\phi} \sin\theta \sin\psi + \dot{\theta} \cos\psi \\ \omega_y &= \dot{\phi} \sin\theta \cos\psi - \dot{\theta} \sin\psi \\ \omega_z &= \dot{\phi} \cos\theta + \dot{\psi}. \end{aligned} \tag{5.9}$$

Note that the components of $\vec{\omega}$ in the expression above are in the body frame. These components must be multiplied by the rotation matrix Eq. 5.8 to generate the equations for the fixed-frame components. The fixed-frame components are clearly more natural for visualization of the rotation, but the equations describing rotational motion are much simpler in terms of the body-frame components. In terms of visualization, it is important to note that the angular velocity in the body frame gives the apparent rotation of the lab frame, except for a change in sign. Thus, for instance, for problems involving constant angular velocity, the body components of the angular velocity are also fixed; not an immediately obvious fact.

If we generate the fixed-frame components of $\vec{\omega}$, we may determine the subsequent motion of the particles constituting the object. This will not, however, give the time development of the Euler angles. This can only be done by solving the differential equations given in Eq. 5.9 for $\phi(t)$, $\theta(t)$, and $\psi(t)$.

The components of the angular momentum, $\vec{L}$, of the rotating object can be written

$$\begin{aligned} L_x &= I_x\omega_x \\ L_y &= I_y\omega_y \\ L_z &= I_z\omega_z. \end{aligned} \tag{5.10}$$

Once again, ω_x, ω_y, ω_z, L_x, L_y, and L_z are the body-frame components of the vectors. To obtain results for the fixed frame, both $\vec{L}$ and $\vec{\omega}$ components must be transformed using Eq. 5.8.

The differential equations of motion for the Euler angles are obtained from the relation

$$\vec{N} = \frac{d\vec{L}}{dt}. \tag{5.11}$$

Transforming this equation into the body coordinates and simplifying gives Euler's equations for the motion of a rigid body subject to an external torque:

$$\begin{aligned} N_x &= I_x\dot{\omega}_x - (I_y - I_z)\omega_y\omega_z \\ N_y &= I_y\dot{\omega}_y - (I_z - I_x)\omega_z\omega_x \\ N_z &= I_z\dot{\omega}_z - (I_x - I_y)\omega_x\omega_y. \end{aligned} \tag{5.12}$$

Note again that the components of the torque and the angular velocity are in the body frame.

5.3 Using the Program

5.3.1 Structure of the Simulation

The ROTATE program allows the user to investigate two types of rotational motion: the motion of an object constrained to rotate about a fixed axis, and the motion of a free object subject to various torques, which may be zero. The dimensions of the rotating object may be changed, thereby changing the moment of inertia. The initial magnitudes and orientation of the angular velocity and torque are adjustable and the initial orientation of the body may also be changed. The object is assumed to consist of equal-mass particles at the eight corners of a rectangular wire-frame, centered at the origin. The dimensions of this object may be changed. The body frame is always taken to be the symmetry axes of this object.

Euler angles are used throughout the simulation to represent the orientation of the object. This requires solution of a differential equation, even for the case of constant angular velocity. For this case a simpler approach would be to generate the rotation matrix for rotations about the fixed axis and apply this to the particles constituting the object. Use of the Euler angles in the constant angular velocity problems, however, is helpful in building an intuitive understanding of the Euler angles. Since the solutions to the differential equations are approximate, however, the orientation of the fixed axis may begin to wander if the simulation is run for a sufficiently long time.

The three-dimensional (3-D) image is produced by rotating the object slightly about the z-axis, then slightly about the y-axis, and then plotting the projection of these points onto the y-z plane. To generate perspective, the y- and z-coordinates of the points are scaled according to the x-coordinate with positive x-values generating an expansion and negative values generating a contraction. This makes objects that are further away appear nearer to each other. Additionally, whenever a line segment lies completely behind the y-z plane it is drawn in a darker shade. The tips of vectors are generated so as to always lie in a plane containing the vector and the z-axis (in the fixed frame). All this should help in visualizing the 3-D system. Occasionally, you will lock onto the image in the wrong perspective and the object will appear to undulate in strange ways as it rotates. When this occurs, simply look away from the screen for a second, then try again.

The stereographic representation is the most fun way of observing the motion. This effect is produced by generating two images with slightly different rotations about the z-axis. If you look at the right image with the right eye and the left image with the left eye, the 3-D sense is remarkably improved, since the slight change in perspective is closer to what you would actually observe if you were looking at a wire-frame model. Looking at a different image with each eye requires some practice. Try turning the monitor brightness down so that you can see your reflection in the screen and then bring the brightness up slowly. Without actually looking at the image (look at your reflection, but *think* about the image) you should see a narrow window in the center of the screen flanked by normal-sized windows on either side. Your task is to increase the size of that central window, by forcing your eyes to look at different images—but don't try to think about what your eyes are doing, until the two images overlap. The lines and vectors should literally jump out of the screen when you get this to happen. Note that having the animation running may be helpful. You might also try including the body-frame screens with the stereographic option since all screens are then smaller and closer together making the stereographic image easier to see.

The spinning top system has been symmetrized by assuming that the torque is applied to the object through two forces acting in opposite directions. The first force is applied vertically downward at a point z_0 on the positive z-axis of the body system while the second force is applied vertically upward at $-z_0$ point on the negative z-axis. The magnitude of the torque is thus

$$\tau = 2Fz_0 \sin\theta . \tag{5.13}$$

The direction of the torque is perpendicular to the plane including the z-axis (body frame) and the z'-axis (fixed frame).

5.3.2 The User Interface

The hot keys at the bottom of the screen allow you to run the simulation to observe the time development of the system. The animation is begun by hitting the F2 key and may be subsequently paused by hitting any key or clicking the mouse. The animation may be speeded up, slowed down, or run in reverse by hitting various hot keys, as indicated.

Different problems may be selected by choosing **Systems** from the main menu. The orientation of the object and initial conditions can all be changed by selecting **Specify Initial Conditions** from the **Parameters** menu. The moment of inertia may also be changed by selecting **Specify Body Dimensions** from the **Parameters** menu. The various menu items in the simulation are described in more detail in the next section. Please note that reasonably detailed context-sensitive help is available at every step while running the program by typing F1.

Program Menus

- **Systems**: This selection allows you to choose one of three systems: rotation about a fixed axis; rotation with a constant torque, which may be set to zero; and motion of a top. Selection of a system resets initial values of quantities to defaults for that system.

- **Parameters**

  - **Initial Conditions**: This selection allows you to set the initial orientation of the object, and the initial values of the angular velocity and the torque (if allowed for this system). The orientation of the object is specified by the three Euler angles. The initial value of the angular momentum is specified by giving two Euler angles (ϕ and θ) and the magnitude of the vector. Note that the third Euler angle has no meaning for a vector. The torque is not available for fixed rotation axis problems and only the magnitude may be changed for the spinning top system. For constant torque problems, the torque is specified in the same manner as the angular velocity. Note that all angles are in degrees here to simplify entering particular fractions of circles (which is, after all, why degrees were invented).

- **Body Dimensions**: This selection allows you to adjust the size of the object. You can click the mouse on the slider arrow to change the dimensions in small increments or click on the bar to produce larger jumps. You may also click and drag the slider for even faster changes, or you may click in the numerical window and enter a specific value. The **arrow** keys may be used to adjust sliders, using the **tab** key to move between sliders. Note that the body dimensions may not be reduced to zero, so that the simulation does not need to deal with the special cases of zero moments of inertia. For the smallest dimensions, however, the object can certainly be visualized as consisting of a simple dumbbell or a flat rectangle.

- **Display**: Selection of this item brings up an input screen that allows you to specify what is drawn on the screen. The motion in the lab frame is always shown. You may elect to display the motion from the body frame as well. You may elect not to include various objects in each screen, to help you concentrate on particular quantities. Note that the torque vector is always displayed as attached to the tip of the angular momentum vector. This is done as a reminder that the torque is responsible for the change in angular momentum and also to help in visualization when many vectors are drawn in one screen. Thus, whenever torque is displayed, the angular momentum vector is also displayed. The scale factors at the bottom of this input screen allow you to adjust the length of the vectors as they are displayed on the screen.

- **Reset**: This item sets the system back to the $t = 0$ state. To reset the default values for a particular system, select that system from the **Systems** menu item.

5.4 Exercises

5.1 **Euler Angles**
Select the fixed rotation axis system for this problem and turn off the display of torque and angular momentum using the **Display** menu choice. Note that for this problem, the Euler angles describing the object will be labeled $(\phi_o, \theta_o, \psi_o)$ and the angles describing the angular velocity will be labeled $(\phi_\omega, \theta_\omega)$.

(a) What values must you give to ϕ_ω and θ_ω to cause the initial angular velocity to point in—

(i) the negative y direction?

(ii) the positive y direction?

(iii) the positive x direction?

Use the simulation to verify your predictions and observe the motion in each case.

(b) Make sure that the initial orientation of the object is the system default orientation: $(\phi_o, \theta_o, \psi_o) = (0°, 30°, 0°)$. When the angular velocity points in the positive x direction, what Euler angle (describing the body orientation) changes when the simulation is run? What happens to all three angles when this angle reaches 180°? Why?

(c) When the angular velocity points in the positive y direction, what Euler angles change when the simulation is run? Run the simulation for $\frac{1}{4}$ of a full period (recall $T = 2\pi/\omega$). What are the values of the Euler angles? Predict the values at $t = T/2$ and $t = 3T/4$; then run the simulation to see if your values are correct.

5.2 **Angular Velocity**

Select the fixed rotation axis system with the default orientations for this problem and turn off the display of torque and angular momentum using the **Display** menu choice. Note that the object precesses about the z'-axis when the simulation is run. Change the orientation of the angular velocity so that it lies along the z-axis of the body, by setting $\theta_\omega = 30°$. Verify that the object now rotates around the body z-axis. We wish to investigate the motion that is generated by an angular velocity that is a sum of these two angular velocities.

(a) If we start with the angular velocity oriented along the z-axis of the body frame and add a small component in the z' direction, you might expect the resulting motion to be the "superposition" of the two individual motions, i.e., the body would rotate about its z-axis, which would slowly precess about the z'-axis. Adding the small z' component to $\vec{\omega}$ causes the angular velocity to tilt toward the z'-axis (the magnitude changes slightly also, but that can be neglected here). Run the simulation with $\theta_\omega = 28°$ to show that the simple superposition notion is not correct. Describe the rotation that does occur (note that it is particularly simple in the body-frame window).

(b) Change the dimensions of the body so that $L_x = 1$ m and $L_y = L_z = 10$ m so that it approximates a flat, square sheet. Change the three Euler angles describing the initial orientation of the body to zero.

(i) Using the relationship $\vec{v} = \vec{\omega} \times \vec{r}$, determine the initial velocities of the points on the four corners of the sheet if the angular velocity is 6.0 rad/s in the z direction. Test your results qualitatively by running the simulation.

(ii) Repeat the calculation in i for the angular velocity equal to 6.0 rad/s in the y direction.

(iii) Use $\vec{\omega} \times \vec{r}$ to determine the initial velocities of these points if $\vec{\omega}$ is the sum of the two angular velocities described above. Show that these are equal to the vector sum of the velocities found in parts i and ii.

(iv) Describe in words how the superposition of the two angular velocities in part iii generates the observed motion of the sheet.

5.3 **Moment of Inertia**

Any two rigid objects with identical moments of inertia will behave exactly the same when subjected to identical torques. Thus, we may replace a complicated rigid object with a simpler representation if the moments of inertia are the same. In this simulation, a rectangular frame is used with a point mass of 1.0 kg at each of the eight corners. The three dimensions of the object determine the three moments of inertia.

(a) If the object has dimensions of 1.0 m in the x and y directions and 8.0 m in the z direction, calculate the moment of inertia about each principal axis. Check your result using **Specify Body Dimensions** under the **Parameters** menu. How different are I_x and I_y from that of the "ideal dumbbell" having zero extent in the x and y dimensions?

(b) Calculate the moment of inertia if the object has dimensions of 10 m in the x and y directions and 1.0 m in the z direction. Use the simulation to check your result.

(c) An alternative simple representation of a rigid body which might be called the 3-D cross consists of six identical masses at symmetric points on the positive and negative coordinates axes. Determine the required positions of 1.0 kg masses to generate the same moment of inertia of part b.

(d) Write down the general formula for I_x, I_y, and I_z for the object in part c and use this to prove the theorem that the sum of any two components of the moment of inertia is always greater than or equal to the third.

5.4 **Angular Momentum for Constant Angular Velocity—I**
Select the fixed rotation axis system with the default orientations and note that the angular momentum is not oriented in the same direction as the angular velocity. Adjust the x and y dimensions of the object to be 1.0 m so that the object approximates an ideal dumbbell.

(a) What angle does the angular momentum make with the axis of the dumbbell? Make a sketch of the dumbbell and show the angular momentum.

(b) Calculate the angular momentum vector of each of the two mass particles (of an ideal dumbbell) using the expression $\vec{L} = \vec{r} \times \vec{v}$, and explain the orientation of the total angular momentum. Remember that $\vec{r}_1$ and $\vec{r}_2$ are relative to the origin of the coordinate system.

(c) Show on your sketch the forces that must act on each mass particle if the system is to rotate at the constant angular velocity. What provides this force?

(d) Determine the direction of the torques generated by the forces investigated in part c. Include these on your sketch and show that their sum agrees with the net torque displayed by the simulation.

5.5 **Angular Momentum for Constant Angular Velocity—II**
Select the fixed rotation axis system with the default orientations and set the x and y dimensions of the object to 10 m and the z dimension of the object to 1.0 m so that the object approximates a thin sheet.

(a) Make a sketch of the object and show the angular momentum vector. How is this result different from that of the ideal dumbbell above?

(b) Using the 3-D cross representation (see Problem 5.3c) of this object, it can be treated as two identical dumbells, one oriented along the x-axis and one along the y-axis. Using results from the previous problem, explain the orientation of the angular momentum vector.

(c) What happens to the angular momentum vector as the object changes from an ideal dumbbell to an ideal flat sheet? Consider in particular the intermediate case when all three dimensions are equal.

5.6 **Constant Torque**

Select the constant torque system and set the initial orientation of the object to $(\phi_o, \theta_o, \psi_o) = (0°, 90°, 0)$. Set the initial angular velocity to zero and the initial torque magnitude to 100 N-m. You might want to change the torque scale factor to make the vector visible on screen.

(a) What orientation of the torque will cause only the θ Euler angle to change? What orientation will cause only the ϕ Euler angle to change? What orientation will cause only the ψ Euler angle to change? For each case, draw a sketch of the object, showing the torque and showing the forces that might be applied to the object to generate this torque.

(b) If the torque is oriented along the z-axis, determine the time required for the object to start from rest and make one full revolution. Run the simulation to check your results.

(c) Suppose the initial angular velocity is 6.0 rad/s oriented along the x-axis, and the torque is 100 N-m oriented along the z-axis. Describe the motion that occurs when the torque is applied. How is it different from the motion in part b? Consider the forces that are applied in generating this torque and try to explain why this motion is different from that in part b. Hint: Consider the velocities and accelerations of the individual mass particles.

References

1. Marion, J. B., Thornton, S. *Classical Dynamics of Particles and Systems*. San Diego: Harcourt Brace Jovanovich, 1988.

6

Collisions

Bruce Hawkins

A hit, a very palpable hit.

William Shakespeare, *Hamlet*

6.1 Introduction

A quick introduction to the features of the computer program is provided in the COLISION walk-through in Appendix A.

A collision is an interaction between two particles in which the forces between the two particles are so large that all other forces may be neglected. Since this would include, for example, binary star systems, we also need to say that it is of limited duration, with the two particles starting out a large distance away from each other and ending up a large distance from each other.

The study of collisions is of interest on a number of levels, ranging from the forensic to the nuclear. At the forensic level, police officers collect evidence to determine who was at fault in an automobile collision. At the nuclear level, physicists use collisions between nuclei or between sub-nuclear particles to determine the properties of the nuclei or the particles. In between, tennis players control the collision between the racket and the ball, and most of us use our eyes to analyze the results of collisions between light photons and the world around us.

The study of collisions can be divided into two parts. One part asks nothing about the forces, using the principles of conservation of momentum and angular momentum. In addition we consider what happens to the energy, which is conserved in some collisions and not in others. These principles are just as applicable in the quantum realm as in the classical. The second part considers the relationship of the force between the particles to their directions of motion after the collision. The usual experimental procedure is to observe the directions and try to deduce the forces. Theory assists by hypothesizing forces and computing the directions.

One of the early notable uses of collisions in physics was Rutherford's discovery of the nucleus. By bombarding a thin gold foil with fast helium nuclei (alpha particles), he found that most of them went right though the foil, but a few of them bounced back. This was as surprising as shooting bullets at tissue paper and having some of them bounce. That would make you suspect that some small, hard, and heavy objects were hidden in or behind the tissue paper. The small, hard, and heavy objects within the gold foil are the nuclei of the gold atoms.

Note this conclusion is based on having many particles go on through in addition to having some bounce. If they all bounced you would just conclude that you had a piece of armor plate disguised as tissue paper. Thus the conclusion depends on observing that different particles come out at different angles: one is observing the *angular distribution* of the particles. From the details of the angular distribution one can draw conclusions about the nature of the forces involved; in Rutherford's case the conclusion was that the forces between the nuclei obey Coulomb's law. The energies he used were high enough that the electrons surrounding the nucleus did not make an observable difference, and low enough that the nuclei did not penetrate each other.

6.2 Theoretical Concepts

The simulation described in this chapter is restricted to central force interactions between two particles. Mechanics textbooks devote a chapter or so to collisions.[1,2] Since the force is central, angular momentum is conserved, and the motion takes place in a plane perpendicular to the angular momentum vector. Although forces in the quantum realm cannot always be described as central, they do still conserve angular momentum. The collision of a pair of particles can therefore be characterized by two parameters which describe the initial relative momentum in that plane.

The first parameter usually used is the energy (or the speed) of one particle, in the coordinate system in which the other is stationary. This determines the magnitude of the momentum. The other input parameter must determine the direction of the momentum, and it is customary to describe it by specifying the closest distance to which the particles would approach each other if there were no forces between them (Fig. 6.1). This distance is called the *impact parameter* and is customarily denoted by the symbol b. This is a natural parameter for collisions of a beam of particles traveling in approximately the same direction towards a stationary target. Particles in the beam *and* in the target are then distributed randomly, and the distribution of impact parameters is readily calculated, as is discussed below, in textbooks,[1,2] and in the tutorial portion of the simulation.

6.2.1 Cross Sections

Physicists, chemists, astronomers, and other scientists are all vitally interested in the probability of different kinds of interaction between two particles resulting from a collision. For example, there may be three different possible outcomes to collisions of the same two particles: one particle may capture the other, kinetic energy may be converted to internal energy of one or both objects, or kinetic energy may

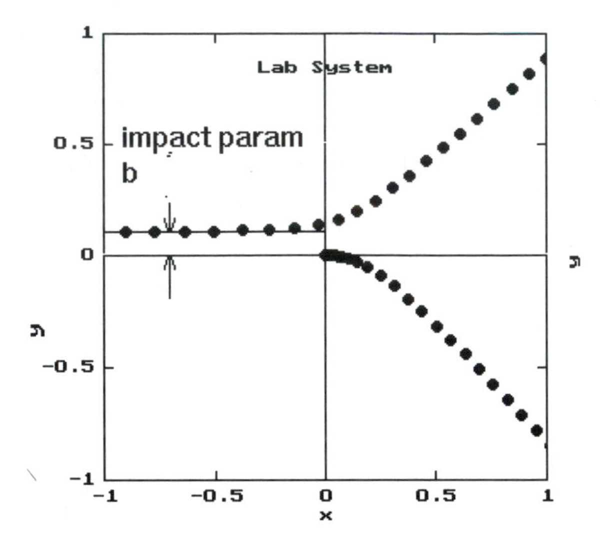

Figure 6.1: Collision showing impact parameter b.

be conserved. The probabilities of the different events are described in terms of an equivalent target area, called a cross section, with units of area.[3] If capture is very likely, then the target area is large, and we say that there is a large capture cross section. Other cross sections of interest are the elastic-scattering cross section, which gives the likelihood of scattering with kinetic energy being conserved; the inelastic-scattering cross section, which tells the probability of kinetic energy being partially converted into internal energy; the cross sections for particular inelastic scattering events, such as the excitation of a particular energy level in an atom or in a nucleus; and cross sections for outgoing particles being different from the ingoing ones, as when a proton hits a heavy nucleus and a neutron emerges, leaving a different nucleus behind.

The usual symbol for cross section is σ.

6.2.2 Differential Cross Sections

All the cross sections described so far are *total cross sections*, summed over all possible impact parameters. In studying the nature of the forces between the particles, which is our main interest here, it is useful to consider *differential cross sections* which describe the angular distribution of particles from similar collisions at different impact parameter. In classical mechanics, a given impact parameter always results in the same scattering angle, that is, the change in direction of the incident particle.

Particle detectors necessarily have a finite size and therefore detect particles scattered through a range of scattering angles, with a corresponding range of impact parameters. Making the range smaller reduces the number of particles detected, and so has a limit. The target area or cross section $d\sigma$ corresponding to a range of impact parameters db at b is shown in Figure 6.2, and is given by

$$d\sigma = 2\pi b \, db \,. \tag{6.1}$$

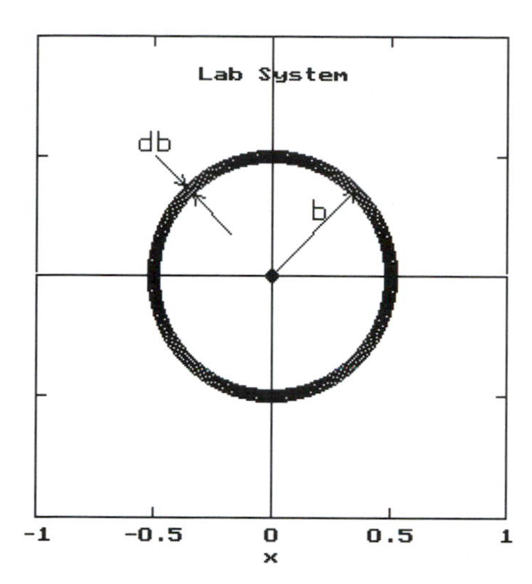

Figure 6.2: Target area depending on a small range of impact parameters db.

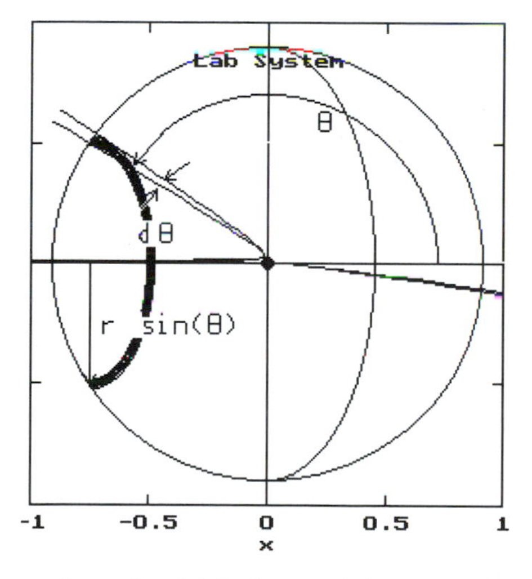

Figure 6.3: Area through which the particles pass after the collision.

The particles striking the ring $d\sigma$ will all be scattered through about the same angle and so will come out in a narrow ring on a sphere surrounding the target, shown in Figure 6.3. The location of the latter ring depends on the forces between the particles. These particles will be more spread out if the area dA' of this latter ring is larger, and that means that fewer of them will strike a detector of a given size, so that the number of particles per unit time striking a given detector of area A' is

$$N' = I\frac{d\sigma}{dA'}A', \tag{6.2}$$

where I is the total number of incident particles per unit area and unit time. N' is the quantity which is measured, and I and A' depend on the particular experiment; $d\sigma/dA'$ is the part that depends on the nature of the collision. We would like to remove the dependence of dA' on the distance r from the target to the detector, which we can do by dividing by r^2, thus transforming dA' into $d\Omega$. The quantity $\Omega = A/r^2$ is called *solid angle*. N' then becomes

$$N' = I\frac{d\sigma}{d\Omega}\Omega\,. \tag{6.3}$$

The area dA' through which the particles emerge is the area of a circle at distance r from the site of the collision; the circumference of the circle is $2\pi r \sin\theta$, and the *width is* $r\theta\, d\theta$. Thus,

$$dA' = 2\pi r^2 \sin\theta\, d\theta \tag{6.4}$$

$$d\Omega = 2\pi \sin\theta\, d\theta\,. \tag{6.5}$$

Dividing Eq. 6.1 by Eq. 6.5, we find

$$\frac{d\sigma}{d\Omega} = \frac{b}{\sin\theta}\left|\frac{db}{d\theta}\right|, \tag{6.6}$$

where the absolute value is used because $db/d\theta$ is usually negative, and the differential cross section (Eq. 6.6) is desired to be positive.

The differential cross section depends upon the forces through $db/d\theta$.[4] This makes the differential cross section a tool in determining what the forces are. You can compare the differential cross sections for different force laws in the simulation (Exercise 9). The tutorial provides assistance in visualizing the meaning of both total and differential cross sections.

6.2.3 Reference Frames

Two frames of reference are of interest: one fixed in the laboratory and one moving with the center of mass. The laboratory system is the one in which the initial conditions of the experiment are controlled. For many years, one of the particles in most nuclear scattering experiments was stationary in the laboratory system. More recently, many experiments are arranged to have both particles moving, usually in such a way that the center of mass is stationary in the laboratory, so that the two systems coincide. Although collision initial conditions are given in the laboratory system, kinematical calculations are easier in the center-of-mass system, enough easier that it is worth performing the simple transformation between the two systems. The tutorial provides assistance in visualizing the transformation and the greater simplicity of the center-of-mass system.

The reason that modern experiments go to the expense of providing a dense beam of moving particles as the "target" (projectile and target become interchangeable in this case) is that the energy expended in making the center of mass move is wasted with regard to exciting internal energy changes or creating new particles, since the momentum of the center of mass cannot change. This is especially true when the energies are high enough to require a relativistic analysis. Although relativity is generally beyond the scope of our treatment, the tutorial does provide a brief introduction to the relativistic center-of-mass transformation.

6.2.4 Computational Approach

The motion of two particles interacting under the action of central forces is simulated using Runge-Kutta fourth-order numerical integration to solve the differential equations of motion. This technique[5] is discussed in chapter 1 and in chapter 4, section 4.2.2. Since motion under the action of central forces is always confined to a plane, the simulation may be restricted to two dimensions without loss of generality.

The potential is calculated only in order to display it; it is not otherwise used, since the motion is calculated by integrating the accelerations.

All simulations described in this chapter use dimensionless units, chosen so as to give a meaningful display on the screen. It is easier to work out what SI units correspond to a given set of simulation parameters than it is to pick an impact parameter that gives a meaningful display for a given energy and force strength.

6.3 *Types of Collisions*

Eight different types of collision are provided in addition to a user-definable type. Three of these are variations on the Coulomb potential, one is an arbitrary power law, two more are nuclear force approximations, and two are macroscopic collisions of hard and soft spheres.

6.3.1 Coulomb Force Variations

The three variations on the Coulomb force are different approximations to the actual forces between charged particles. The pure Coulomb potential and force,

$$V_c(r) = \frac{k}{r} \tag{6.7}$$

$$F_c(r) = \frac{k}{r^2}, \tag{6.8}$$

apply to interactions between electrons and between electrons and protons at low energies; k will be negative for attractive interactions.

For interactions between atomic nuclei, the Coulomb force is a good approximation if the collision energies are high enough to make the interaction with the electrons surrounding the nucleus negligible and low enough that the nuclei do not interpenetrate. If the energies are low, a shielded force taking into account the interaction with the electrons is appropriate, while if the energies are high, then the simplest approximation is to keep the force constant (truncated force) for distances less than some radius at which nuclear forces are assumed to come into play. While one can combine the Coulomb force with a nuclear force, this results in a differential equation which is difficult to integrate and still maintain precision.

The shielded approximation multiplies the Coulomb $V_c(r)$ by an exponential factor,[6] making the force fall off more rapidly:

$$V_s(r) = V_c e^{-r/r_0} \tag{6.9}$$

$$F_s(r) = F_c\left(1 + \frac{r}{r_0}\right)e^{-r/r_0} \tag{6.10}$$

to approximate the shielding of the nuclear charge by the surrounding electrons.

The truncated force law is

$$F_s(r) = \begin{cases} F_c, & r > r_0 \\ \frac{k}{r_0^2}, & r \leq r_0 . \end{cases} \tag{6.11}$$

6.3.2 Nuclear Forces

Two force models between nuclei are offered: Yukawa and Woods-Saxon. The Yukawa potential,[7]

$$V_Y(r) = -V_0 \frac{e^{-r/c}}{r}, \tag{6.12}$$

was proposed by Yukawa in the 1930s when he suggested that nuclear forces were mediated by then-unknown particles of masses intermediate between the electron and proton, which have come to be known as mesons. It is a reasonable approximation to the force between two single nucleons. Using

$$F(r) = -dV/dr, \tag{6.13}$$

the force is

$$F_Y(r) = -V_0 e^{-r/a}\left(\frac{1}{r^2} + \frac{1}{ar}\right). \tag{6.14}$$

The Woods-Saxon potential and force,[8]

$$V_{WS}(r) = -V_0 \frac{1}{1 + e^{(r-R_0)/a}} \tag{6.15}$$

$$F_{WS}(r) = -V_0 \frac{V_0 e^{(r-R_0)/a}}{a(1 + e^{(r-R_0)/a})^2}, \tag{6.16}$$

is a better approximation for nuclei composed of many nucleons where the potential is relatively constant within the nucleus. No exact expression is known for the forces between nuclei or the particles composing them, and the interactions are fundamentally quantum mechanical in nature, so that we are dealing here with semi-classical approximations to the actual forces.

6.3.3 Hard Sphere Collisions

This simulates the collision of two non-rotating billiard balls. The velocity component perpendicular to the contact surfaces is reversed at the time of contact. Exercise 9 examines the effect of considering rotation.

6.3.4 Soft Sphere Collisions

The force is assumed to be a spring force along the line of centers of two identical spheres when their centers are closer than twice the radius of a sphere:

$$V(r) = \begin{cases} \frac{1}{2}kr^2, & r < R \\ 0, & r \geq R. \end{cases} \tag{6.17}$$

6.4 Using the Program

6.4.1 Setting a System Into Motion

The simplest way to start a collision is to click the mouse button with the cursor in the **Lab System** window. The user can also select either the F2–**Go** hotkey or the F3–**NewData** hotkey. **Go** repeats the collision most recently displayed, or, when a new force has just been chosen, a collision with the default parameters for that force. F3–**NewData** presents a **Collision Type** dialog box offering a choice between a single collision or a set of collisions with a systematic variation of a parameter chosen from this dialog box. F3–**NewData** does not clear the screen in order to make possible comparisons between different collisions.

If the user chooses the single collision, a **Collision Conditions** dialog box appears offering the opportunity to change any of the relevant parameters. After all desired choices are made, the collision simulation takes place.

Alternatively, the user can select an impact parameter by clicking the mouse with the cursor in the **Lab System** window. The numerical value of the impact parameter chosen will be displayed.

If the user chooses **Vary ....**, a dialog screen will appear with the opportunity to change the default values for the range of variation and the number of collisions to be displayed. A check box is also offered that brings up the **Collision Conditions** dialog box. The requested number of collisions will then be displayed without clearing the screen between collisions.

Vary Impact Parameter and **Plot Cross Section** differ only in that the latter displays a plot of the differential cross sections in place of the center of mass system trajectories. In either case, the dialog box that appears asks whether the user wants the impact parameters spread evenly across the range, or more heavily concentrated near zero. Depending on the collision force law, one or the other will give more information.

The box for the **Cross Section** choice also allows you to save the graph. If you choose to save it, you will be asked for a description. You can later display any or all saved cross sections, identified by their descriptions. The logarithm of the cross section is also displayed, because many cross sections vary too rapidly to display well. All cross sections are calculated relative to the largest cross section observed, usually the one at the smallest scattering angle.

Things to Notice

The criterion for stopping the simulation after a collision is that either object reaches a distance of three times the size of the window. It is not possible to require *both* objects to leave the window because in some collisions one object stops dead. For this reason, you will sometimes see only a short track for one object or the other.

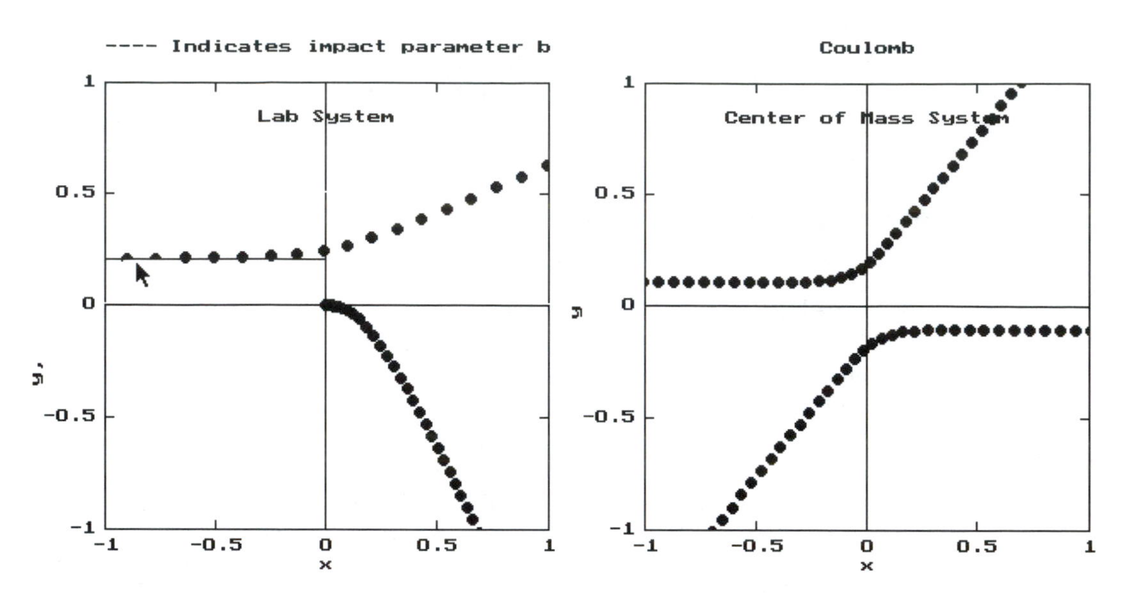

Figure 6.4: Laboratory and center-of-mass trajectories, showing object history.

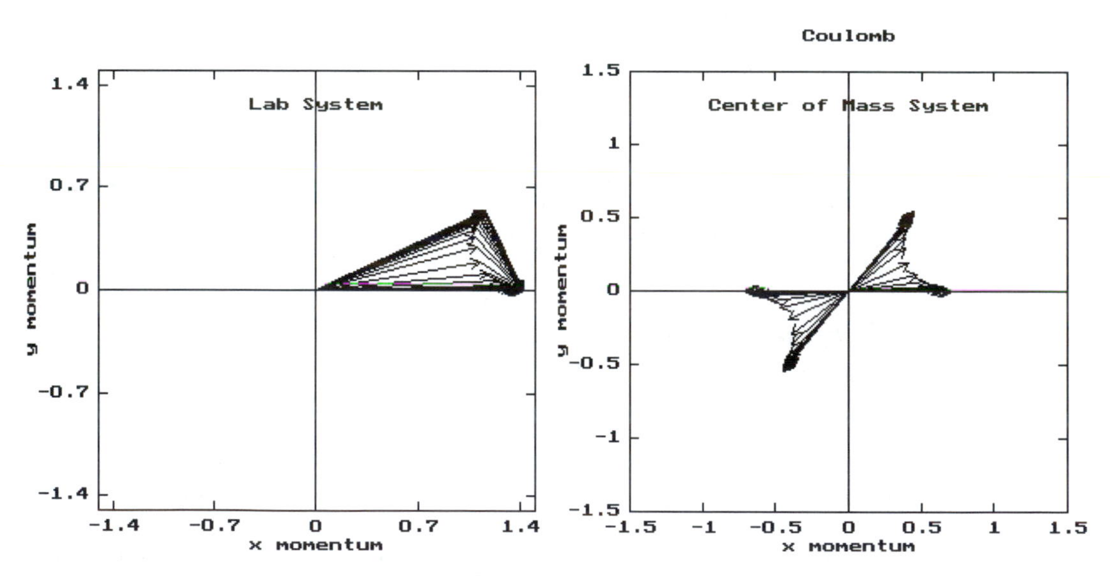

Figure 6.5: Laboratory and center-of-mass momentum vectors, showing object history.

The cross sections as first calculated are not always single-valued functions. Exercise 5 explores the consequences.

6.4.2 Graphical Display

The default screen (Fig. 6.4) shows the trajectories of the particles in the laboratory system on the left, and in the center-of-mass system on the right. The user can choose to display instead the momentum vectors in the two systems (Fig. 6.5), to display all four diagrams simultaneously, or to display any single diagram.

6.4.3 Changing Parameters—Collision Conditions

The **Collision Conditions** dialog box appears when a single collision is chosen or when a range is chosen and the box to change other parameters is checked. This dialog allows the user to set values for the impact parameter, initial distance between particles, mass ratio, energy of the incident particle, and all parameters of the force.

The default initial position of the incident particle is well off the screen for most forces, because many of the forces are long-range, and the collision would not be realistic if the initial position were well within the range of the force. On the other hand, if the display scale were enlarged to make the initial position on-screen, interesting details would be invisible. You can change the scale using **Zoom.**

6.4.4 Display Choices

In addition to changing the parameters, the **Collision Conditions** dialog box allows the user to choose how trajectories are displayed. The choices are **Show Object Now**, which displays two filled circles showing the current positions of the objects, giving you a movie of the collision; **Show Object History**, which makes each object leave multiple pictures of itself or its vectors behind as shown in Figure 6.4 and Figure 6.5, tracing out its path; and **Show Paths Only**, which just draws a line showing the trajectory of each particle.

An additional choice is whether to display the potential or the effective potential. The concept of effective potential is explained in Exercise 7. The potential energy graph is drawn in the **Laboratory System Trajectory** window.

6.4.5 Program Menus

This section describes the menu options found in the program. They may be selected (as discussed in chapter 1) with the mouse; or by pressing F10, **arrow** keys, and the **Enter** key.

- **Files**: Allows access to special files that can be created and read by the program, information about the program and about CUPS, and exit from the program.
  - **About Collisions** and **About CUPS**: Brief descriptions of program and of the CUPS project.
  - **Configure**: Allows you to change the path to the directory in which cross section and history files are stored (shortcut key **Alt-d**), change the display colors (shortcut key **Alt-c**), or check the amount of memory available to the program (shortcut key **Alt-m**). You can change any of the sixteen colors into any other color. In particular, you can easily change to black-and-white, suitable for printing. The procedures for printing are described in chapter 1.
  - **Read Cross Section File**: Read and display a previously saved file containing the cross sections calculated for a series of collisions.
  - **Save Cross Section File**: Save the cross section data now in memory, which may contain data on up to ten collision series with different forces or

values of force constants. The collision descriptions are saved and serve to identify the file. It is an ASCII file.

- **Read Collision History File**: Read and display a previously saved file containing the collision histories for one or more collisions.
- **Save Collision History File**: Save data on the most recent collisions produced from the F3–**New Data** hotkey. The data saved include the nature of the force, force constants, impact parameters, energies, closest approach, and scattering angle, as well as the positions and velocities at each time step. You will be asked for a description to identify the collision when you examine the file later. It is an ASCII file with comments.
- **Exit Program**: This can also be accomplished by **Alt-X**.

• **Plots & Zoom**: The choices on this menu control what is displayed and how it is shown.

- **Choose One Window**: Allows any one window of those described below to fill the entire screen.
- **Trajectories Only**: Displays two windows showing the particle trajectories in the laboratory and center-of-mass systems. This is the default display.
- **Vectors Only**: Displays two windows showing the momentum vectors in the two systems.
- **Trajectories & Vectors**: Displays all four of the above windows at once. It will never be used when a parameter is varied: **Trajectories Only** will be restored.
- **Zoom In** and **Zoom Out**: Change the scale of the windows displayed.
- **Default Scales**: For use if you have zoomed out and in until you are thoroughly confused.
- **Display Saved Cross Sections**: Each time a cross section graph is calculated, the **Cross Section Parameters** dialog box gives you an opportunity to choose **Save**. (**Don't Save** is the default.) If you choose to save the graph, you will be asked for a brief description, which you need to identify this graph when you ask to have the graphs displayed. Up to ten different graphs can be saved, and you can ask to display all of them or some of them to compare with each other. You also have the choice of two different displays. These graphs will differ in their vertical scale from those made during the collision, since it is possible to make a better choice of scale after all the values are known, and necessary when there are several collisions. The set of graphs can be saved to a file as described above.

• **Forces**: This is where you choose the force law. The force laws are described in section 6.3. The choices are—

- **Pure Coulomb**: Eq. 6.8.
- **Coulomb With Shielding**: Eq. 6.10.

- **Coulomb Truncated Near r = 0**: Eq. 6.11.
- **Other Power Law**: Exponent(s) of your choice.
- **Hard Sphere**: Velocity perpendicular to contact surface reverses.
- **Soft Sphere**: Eq. 6.17.
- **Yukawa**: Eq. 6.14.
- **Woods-Saxon**: Eq. 6.16.
- **User-Defined**: To make this choice active, you need to add code to three subroutines: **User_Accel, User_Potential,** and **User_Parameters,** all of which are in COLPHY.PAS. Comments are provided in each to tell you what code is needed.
- **Previous Force**: This allows you to switch back and forth between two different force laws, preserving all constants and initial conditions, without clearing the screen, to facilitate comparisons between the two.

• **HELP!!**: This menu offers two choices:

- **Help**: Access to the help screens, one for each menu item.
- **Tutorial**: Provides text and graphical explanations of cross section topics and center-of-mass system topics.

6.5 Exercises

6.1 **Units: Coulomb Force**
The units used in the simulation are dimensionless and the values of the constants in the force laws are chosen to yield a good range of scattering angles rather than to correspond to any particular situation in nature. Thus the equation used for the Coulomb acceleration is

$$\mathbf{a} = \frac{k\mathbf{r}}{mr^3}. \tag{6.18}$$

(a) What is the value of k/m in SI units for an electron moving in the field of a proton?

(b) What are the units of k/m in the SI system?

(c) Let an electron speed of 100 m/s correspond to 1 simulation unit. The default value of k/m for the Coulomb force simulation is 0.1; assume that this corresponds to an electron striking a proton. Since the units of v and of k/m both depend only on length and time, what must one simulation time unit be in seconds? One simulation length unit in meters? [Hint: 100 m/s = 1 l/t, where l and t stand for the unknown simulation length and time units. Write a similar equation for k/m, and solve the two equations simultaneously for l and t in terms of m and s.

(d) Now assume that v m/s corresponds to v_s l/t, and k/m in SI units corresponds to k_s/m_s in the simulation system. Write the general expressions for l and t.

(e) What are the length and time scale conversion factors to make the default simulation collision diagram ($k/m = 0.1\ l/t$, $v = 1\ l/t$) represent an alpha particle (helium nucleus) incident on an iron atom with four electrons missing (Fe^{++++})?

6.2 **Units: Woods-Saxon Force**

Repeat Exercise 1 for the Woods-Saxon force model. Look to see what the constant values used in the simulation are (in the **Collision Conditions** dialog box) and look up the range and depth of the potential well for a nucleus of your choice to get the appropriate numerical values.

6.3 **Cross Section Calculation**

Save and print out a cross section output file for a pure Coulomb collision. Assume that this corresponds to electrons scattering from protons.

(a) Derive an equation for $db/d\theta$ for the Coulomb force law. If you need help, that can be found in many books such as Marion.[9] Substitute your result into (Eq. 6.6) to obtain an expression for the differential cross section.

(b) For each pair of points in the file, calculate the cross section in simulation units.

(c) Plot the results on a graph, and plot the results of part a on the same graph, multiplied by a factor chosen to make the two graphs coincide at at least one angle.

(d) Calculate the cross sections in SI units.

(e) Calculate the cross sections for the same conditions from the equations in your textbook. Do they correspond?

6.4 **MultiValued Cross Sections**

Save and print out a cross section file for the Woods-Saxon interaction with the default parameters, which appears to be a multi–valued function. Is this acceptable? Why does it happen? What do we do about it?

(a) What will a detector placed at an angle where the function appears to be multi–valued see? One branch (which one)? Both? What we want to calculate is what the detector sees.

(b) What is the largest scattering angle present in the data?

(c) For what impact parameter does this occur?

(d) Calculate some differential cross sections for impact parameters ranging from a few below part b to a few above.

(e) State in words why the raw function is multivalued.

(f) State your conclusion of what should be done about it.

6.5 **Attractive vs. Repulsive Interactions**
Use the knowledge you have gained from Exercise 4 in the following:

(a) Change one of the default parameters in the Woods-Saxon interaction to make the interaction repulsive. You will know you have succeeded when the potential becomes a hill instead of a hole. Which parameter did you change?

(b) Make sure all the other parameters have the default values, and that the one you changed has the same magnitude that it did before. What is the difference between a collison with impact parameter b near zero with energy 0.5 and energy 1.5? At what energy does the change from one regime to the other occur?

(c) Plot cross sections for energies 0.5 and 1.5. Describe the difference between them. Explain the difference in light of your conclusions from Exercise 4.

6.6 **Closest Approach**

(a) Calculate the distance of closest approach for the default Coulomb simulation collision diagram ($k/m = 0.1\ l/t$, $v = 1\ l/t$) for $b = 0$ and $b = 0.1$. Does this agree with what you see on the diagram?

(b) What is this distance for each of the cases in Exercise 1c and e?

6.7 **Effective Potential**
Some textbooks discuss the concept of effective potential.[10] To summarize it briefly, the energy equation for central force motion includes a term which depends upon the angular momentum and otherwise only on r. For constant angular momentum, which is the case for any given collision, this term can be combined with the potential energy term which also depends only upon r. The combination is called the effective potential.

(a) Make a table of the values of the effective potential for various radii for the cases defined in Exercise 6, and for $b = 0.5$. Use values of r ranging from 0.0 to 1.0.

(b) Graph the tables you have just made. Compare them to the graphs generated by the computer when you select **Effective Potential** in the **Collision Conditions** dialog box.

(c) Describe in words the difference between the effective potential graph for $b = 0$ and the others. Why would you expect this difference?

(d) What happens to the minimum in the effective potential as b increases? Does it stay in the same place or move in which direction? Why would you expect this to happen?

6.8 **Shielding**
Consider a proton incident on a doubly ionized calcium atom (Ca^{++}).

(a) Use the methods of Exercise 1 to establish the correspondence between simulation units and SI units for this case.

(b) Now consult a reference such as Jackson[6] and calculate the value of r_0 for the shielding equation given in section 6.3.1.

(c) Save the cross section graph for this shielded Coulomb force and compare it to the saved cross section graph without shielding. Describe the difference and explain it.

6.9 **Comparing Effects of Forces**

In what ways do any two of the force laws differ? The facilities offered by the program will suggest things that you can explore. A large part of this exercise is your own ability to plan an exploration and to notice significant events as you go along.

This is capable of extension to more forces if you wish or if your instructor assigns you to do so.

6.10 **Another Truncation Expression** (Programming Project)

Jackson[11] truncates the potential in a different way:

$$V_s(r) = \begin{cases} V_c, & r > r_0 \\ \frac{3}{2}\frac{V_0}{r_0}\left(1 - \frac{r^2}{3r_0^2}\right), & r \leq r_0 . \end{cases} \tag{6.19}$$

Derive the corresponding force equation and modify **User_Accel** and **User_Potential**, and **User_Parameters** in COLPHY.PAS. Compare the results with the results of the **Coulomb Truncated** menu choice. How do they differ?

6.11 **Realistic Billiard Balls** (Programming Project)

Copy the contents of the **Hard Sphere** procedures to **User_Accel** and **User_Parameters** in COLPHY.PAS and modify them to include rotation.

Ignore the rolling of the balls on the table and consider only the rotation resulting from the contact between the balls. Assume that the collision makes the relative speed of the two points that contact each other zero, and that this happens instantaneously, with the change in rotational angular momentum of the two balls being equal and opposite. Assume that the total energy is conserved, so that the linear kinetic energy of the balls is reduced to allow for the rotational kinetic energy that is acquired. Describe how the resulting collisions differ from hard sphere collisions without rotation.

6.12 **More Realistic Billiard Balls** (Programming Project)

In Exercise 11, also take into account the interaction with the table, assuming that each ball rotates about a horizontal axis like a wheel. Assume that the rotational changes happen instantaneously. How does this change the collisions?

6.13 **Still More Realistic Billiard Balls** (Programming Project)

In Exercise 12, discard the assumption that the change happens instantaneously. Assume that the ball slides on the table while friction is changing its rotation about the horizontal axis. Thus energy will be lost to friction. Wallace and Schroeder[12] and Onoda[13] discuss this problem. How does this change the collisions?

6.14 **Combined Forces: Integration Methods**

(a) Calculate the appropriate constant values for combined Coulomb and Woods-Saxon forces for a proton colliding with a proton.

(b) Do you expect this to make a difference for large impact parameters or for small? Explain why.

(c) Use **Previous Force** to switch back and forth between this and pure Coulomb and explore the region of b that you expect to make a difference. At what b do you see a difference in the scattering angle?

(d) You may find in part c that the simulation for the combined force crashes with the error message that too many iterations have been required to achieve the specified precision. A crude way to deal with this is to reduce the precision demanded. Do you find that you can get results without reducing the precision so far that the simulation becomes meaningless? You can judge this by running the **Pure Coulomb** simulation with the same precision and seeing how much this changes the scattering angle.

(e) (Optional and Difficult: A Major Project). A better way to handle the problem in part d is to use an integration method better suited to this kind of problem. Different integration methods are suited to different kinds of problems, and none are suited for all. The combination of the two forces results in what is known as a *stiff* differential equation. Consult books on numerical analysis, pick a different integration method, and write a new integration routine that you can substitute for **StepRKF.**

References

1. Baierlein, R. *Newtonian Dynamics.* New York: McGraw-Hill, 1983.

2. Fowles, G. R. *Analytical Mechanics.* New York: Saunders College Publishing, 1986.

3. Baierlein, *op. cit.*, p. 286ff.

4. Marion, J. B. *Classical Dynamics of Particles and Systems.* New York: Academic Press, 1965, p. 327ff.

5. Press, W. H. et al. *Numerical Recipes.* Cambridge: Cambridge University Press, p. 550ff, 1986.

6. Jackson, J. D. *Classical Electrodynamics.* New York: John Wiley & Sons, p. 453, 1962.

7. Segre, E. *Nuclei and Particles; An Introduction to Nuclear and Subnuclear Physics*. New York: W. A. Benjamin, p. 615, 1964.

8. Segre, *op. cit.*, p. 467.

9. Marion, *op. cit.*, p. 332ff.

10. Marion, *op. cit.*, p. 276f.

11. Jackson, *op. cit.*, p. 454.

12. Wallace, R. E., Schroeder, M. C. Analysis of billiard ball collisions in two dimensions. American Journal of Physics **56:**815–819, 1988.

13. Onoda, G. Y. Discussion of "Analysis of billiard ball collisions in two dimensions." American Journal of Physics **57:**476–478, 1989.

7

Coupled Oscillators in One and Two Dimensions

Randall S. Jones

7.1 Introduction

To obtain a quick introduction to this simulation see the COUPOSC walk-through in Appendix A.

This chapter deals with the analysis of two or more objects interacting through harmonic forces, that is, two-body forces that are proportional to the relative separations of the two objects. The standard model for coupled oscillators is a set of objects connected by springs. Generalizing from an understanding of the behavior of a single object subject to a harmonic force, we will find that several different vibration frequencies or "natural modes" are possible in which these systems oscillate in a strictly periodic fashion. The motion associated with these individual modes can be combined to generate more complicated motions. Conversely, the complicated motions of a harmonic system can be "broken down" into a linear combination of periodic motions, providing an important tool for understanding the behavior of these systems.

It would be difficult to overemphasize the importance of coupled oscillators in applications of classical mechanics. The subject has applications that range from chemical analysis of molecular vibrations, to the study of electric circuits, to the study of vibrations of mechanical structures. Any system that has a well-defined equilibrium will be subject to some sort of oscillations about that equilibrium. If the oscillations are sufficiently small, the methods discussed in this chapter will most surely find application. If the oscillations are not small, then nonlinear effects may be important and the methods discussed in chapter 4 on anharmonic systems must be applied. Even in these cases, however, the harmonic approximation to the real system will often provide important insights into the behavior of the system. The exercises at the end of this chapter introduce some of the many applications of coupled oscillators.

This simulation will help you investigate the behavior of a wide range of harmonic systems. Given a set of objects and springs connected in one or two dimensions, the simulation can "solve the problem" by generating the normal-mode frequencies and their corresponding motions. It takes any set of initial conditions and resolves them into their component normal-mode motions, or it takes any set of initial mode amplitudes and displays the corresponding motions of the objects. It can also determine the motion of the system when it is acted on by external forces. In this case the total forces are no longer harmonic, so the solution cannot be worked out analytically, but the harmonic analysis gives us an important tool for investigating and understanding the subsequent motion.

7.2 Theoretical Concepts

The theory of coupled oscillators is developed in every standard classical mechanics textbook. Most textbooks use the Lagrangian formulation of mechanics in this derivation, whereas the discussion below uses standard Newtonian mechanics. In the first section we consider a system of objects in one dimension connected by Hooke's-law springs. The extension to two dimensions is given at the end of the section. Textbooks usually take a more general approach by considering any system whose potential energy can be written as a quadratic function of the system coordinates. This allows treatment of problems such as coupled pendula. Suggestions are given for modifications to the program to include such examples. An alternative for systems with three or fewer degrees of freedom is to use the Motion Generator simulation described in chapter 2.

7.2.1 One-Dimensional Coupled Oscillators

Consider a set of N objects, each with mass m_i at equilibrium position x_i^0. The equilibrium distance between object i and object j is

$$L_{ij}^0 = |x_j^0 - x_i^0| . \tag{7.1}$$

If objects i and j are connected by a spring with equilibrium length L_{ij}^0 and spring constant K_{ij}, then a displacement of either object away from its equilibrium position will result in a force acting on each object. If the displacement of object i away from its equilibrium position is given by Δx_i, then the force acting on object i due to spring K_{ij} is given by

$$F_i = -K_{ij}(\Delta x_i - \Delta x_j) . \tag{7.2}$$

The total force on object i due to all the springs is thus

$$F_i = -\sum_{j\neq i}^{N} K_{ij}(\Delta x_i - \Delta x_j) = \sum_{j\neq i}^{N} K_{ij}\Delta x_j - \left[\sum_{j\neq i}^{N} K_{ij}\right]\Delta x_i . \tag{7.3}$$

Note that if there is no spring connecting a particular pair of objects, then this value of K_{ij} is zero.

If object i is connected to a rigid support by a spring with spring constant K_{is}, then another term of the form $-K_{is}\Delta x_i$ is added to the equation above. The following notation can be used to represent additional terms from rigid supports:

$$F_i = \sum_{j \neq i}^{N} K_{ij} \Delta x_j - \left[\sum_{j \neq i}^{N+S} K_{ij} \right] \Delta x_i , \tag{7.4}$$

where S is the total number of rigid supports. From Eq. 7.4 we see that the force that acts on object i will always be a linear function of the Δx_j's. A matrix representation is then convenient. If we define an N-component vector, $\underline{\Delta x}$, to represent the N displacements and another N-component vector, $\underline{F}$, to represent the forces on the N objects,

$$\underline{\Delta x} = \begin{bmatrix} \Delta x_1 \\ \Delta x_2 \\ \vdots \end{bmatrix} \qquad \underline{F} = \begin{bmatrix} F_1 \\ F_2 \\ \vdots \end{bmatrix}, \tag{7.5}$$

then the expression for the force can be written

$$\underline{F} = -\underline{\underline{\mathcal{D}}} \cdot \underline{\Delta x}, \tag{7.6}$$

where $\underline{\underline{\mathcal{D}}}$ is called the *dynamical matrix*. The individual components of $\underline{\underline{\mathcal{D}}}$ are

$$\mathcal{D}_{ij} = \begin{cases} -K_{ij} & i \neq j \\ \Sigma_{k \neq i}^{N+S} K_{ik} & i = j . \end{cases} \tag{7.7}$$

The equation of motion can be written for object i as

$$m_i \Delta \ddot{x}_i = F_i = -\sum_{j=1}^{N} \mathcal{D}_{ij} \cdot \Delta x_j . \tag{7.8}$$

It will be convenient to incorporate the mass of the object into the dynamical matrix by defining a generalized coordinate $q_i = \Delta x_i \sqrt{m_i}$. Then, if we define

$$\tilde{\mathcal{D}}_{ij} = \mathcal{D}_{ij} / \sqrt{m_i m_j} , \tag{7.9}$$

the equation of motion becomes

$$\sum_{j=1}^{N} \tilde{\mathcal{D}}_{ij} \cdot q_j = -\ddot{q}_i . \tag{7.10}$$

As is the case with most ordinary differential equations, we now make a guess at the form of the solutions. Explicitly making use of our intuitive notion that the motion of this system will be oscillatory, we try

$$q_i(t) = A_i \cos(\omega t + \phi) \tag{7.11}$$

for each object. This expression implies that all the objects oscillate with the same frequency and phase, but with different amplitudes. This cannot be the whole story, of course, since the motion of coupled oscillators is more complicated than that. We will find that there are actually N solutions with the form given above. It will then be linear combinations of these solutions that generate the full solution to the problem.

Our initial guess (Eq. 7.11) appears to have $N + 2$ arbitrary parameters ($N A_i$'s, ω, and ϕ), but on substitution into the equation of motion, we will find that this can only satisfy the differential equation if the frequency has a specific value and if the A_i have specific values except for an overall scale factor. Thus, this guess will contain only two free parameters: the overall scale factor and the phase. Fortunately, our analysis will turn up N different allowed frequencies, each with its own set of specific values for the amplitudes, A_i. Thus $2N$ free parameters are obtained, which is exactly what we need in order to specify the initial displacements and velocities of the system. Each solution for a specific value of ω is called a *normal mode* of the system, and the overall scale factor for the amplitudes is often referred to as the *occupation* of that mode.

Let us proceed by substituting Eq. 7.11 for the q_i into the equations of motion:

$$\sum_{j=1}^{N} \tilde{\mathfrak{D}}_{ij} \cdot A_j \cos(\omega t + \phi) = -\frac{d^2}{dt^2} A_i \cos(\omega t + \phi) = \omega^2 A_i \cos(\omega t + \phi) . \quad (7.12)$$

$\cos(\omega t + \phi)$ appears on both sides of the equation and can be eliminated. The matrix equation is now an equation for the A_i's:

$$\sum_{j=1}^{N} \tilde{\mathfrak{D}}_{ij} \cdot A_j = \omega^2 A_i . \quad (7.13)$$

This is an example of an eigenvalue equation. There will be N solutions, each corresponding to a particular value of ω (ω^2 is called the *eigenvalue*, ω is called the *normal mode frequency*) and a particular set of values of the A_i (called the *eigenvector*, or *normal mode*). It is clear from Eq. 7.13 that the A_i's are defined only up to an overall scale factor. The specific normal mode frequencies are labeled ω_n with n running from 1 to N. For each ω_n, the amplitudes are labeled by A_i^n. It is important to remember that the i subscript in this notation corresponds to the particular object, while the n superscript corresponds to the particular normal mode. Individual modes will often be labeled with letters rather than numbers (i.e., ω_a, ω_b, etc.)

The expression representing the motion of object i is written as a sum of the normal modes:

$$\Delta x_i(t) = q_i/\sqrt{m_i} = \frac{1}{\sqrt{m_i}} \sum_{n=1}^{N} A_i^n \cos(\omega_n t + \phi_n) . \quad (7.14)$$

Sometimes, in order to make the free parameters more obvious, this expression is written as

$$\Delta x_i(t) = \frac{1}{\sqrt{m_i}} \sum_{n=1}^{N} \lambda_n \tilde{A}_i^n \cos(\omega_n t + \phi_n) , \quad (7.15)$$

and the arbitrary scale factors for the $\tilde{A}_i^n$ (for each n) are fixed by requiring that

$$\sum_{i=1}^{N} [\tilde{A}_i^n]^2 = 1 . \quad (7.16)$$

Note that in Eq. 7.15 the values of the $\tilde{A}_i^n$ and ω_n are fixed by the physical parameters of the system, while λ_n and ϕ_n are parameters used to match the initial conditions of the system. This is very similar to the single oscillator case in which the oscillation frequency is fixed by the physical parameters of the system (K and m), while the phase and the amplitude determine the initial state of the oscillator.

If only one scale factor is non-zero, all objects oscillate with identical frequency and phase. The fact that the motion of any coupled oscillator (in the harmonic approximation) can be described as a linear combination of such motions is a very important tool for analyzing the motion of complex systems. Understanding the nature of these modes and how they combine to generate more complex motion is very important and it is the primary goal of this simulation.

The parameter λ_n is directly related to the energy stored in this mode. To understand this, consider the velocity of object i when a single mode λ_n is non-zero:

$$\Delta \dot{x}_i(t) = -\frac{1}{\sqrt{m_i}} \lambda_n \tilde{A}_i^n \omega_n \sin(\omega_n t + \phi). \tag{7.17}$$

The maximum velocity is just

$$\Delta \dot{x}_i^{max} = \times \frac{\lambda_n \tilde{A}_i^n \omega_n}{\sqrt{m_i}}, \tag{7.18}$$

so the maximum kinetic energy is given by

$$T_i^{max} = \frac{1}{2} m_i \dot{x}_i^2 = \frac{1}{2} \lambda_n^2 (\tilde{A}_i^n)^2 \omega_n^2. \tag{7.19}$$

Since only one mode is present, all the objects reach their maximum velocities simultaneously (since they are oscillating in phase) when the displacements (and hence the spring potential energies) are zero. Thus, the total energy of the system is just the sum of the maximum kinetic energies of each particle,

$$E_{tot} = \sum_{i=1}^{N} T_i^{max} = \frac{1}{2} \lambda_n^2 \omega_n^2 \sum_{i=1}^{N} (\tilde{A}_i^n)^2 = \frac{1}{2} \lambda_n^2 \omega_n^2, \tag{7.20}$$

using the normalization relation Eq. 7.16.

The determination of the eigenvalues and eigenvectors of a particular dynamical matrix is fairly straightforward. For a small number of degrees of freedom the solution can be worked out directly (see Arfken,[2] for example). For larger matrices, various numerical methods are available to determine eigenvalues and eigenvectors to whatever level of accuracy is required. The present simulation uses the Jacobi method to determine the quantities. Descriptions of the method can be found in any standard numerical computing text. The version used has been adapted from *Numerical Recipes*.[1]

7.2.2 Two-Dimensional Coupled Oscillators

If the objects are free to move in two dimensions, the motions are a little harder to visualize, but the analysis is not much more complicated than the one-dimensional case.

Consider a set of N objects each with mass m_i at equilibrium position given by

$$\vec{r}_i^{\,0} = x_i^0 \hat{e}_x + y_i^0 \hat{e}_y. \tag{7.21}$$

The equilibrium distance between objects i and j is

$$L_{ij}^0 = |\vec{r}_j^{\,0} - \vec{r}_i^{\,0}| = \sqrt{(x_j^0 - x_i^0)^2 + (y_j^0 - y_i^0)^2}. \tag{7.22}$$

If objects i and j are connected by a spring with equilibrium length L_{ij}^0 and spring constant K_{ij}, then a displacement of either object away from its equilibrium position will result in a force acting on each object. If the current position of object i is given by

$$\vec{r}_i = (x_i^0 + \Delta x_i)\hat{e}_x + (y_i^0 + \Delta y_i)\hat{e}_y, \tag{7.23}$$

and the current distance between these objects is written as

$$L_{ij} = |\vec{r}_j - \vec{r}_i|, \tag{7.24}$$

then the magnitude of the force is given by

$$F_{ij} = K_{ij}|L_{ij} - L_{ij}^0|. \tag{7.25}$$

The direction can be incorporated into this result by multiplying by the unit vector that points from object i to object j:

$$\hat{e}_{ij} = \frac{\vec{r}_j - \vec{r}_i}{L_{ij}}. \tag{7.26}$$

The vector force that the spring exerts on object i can thus be written

$$\vec{F}_i = K(L_{ij} - L_{ij}^0)\hat{e}_{ij}. \tag{7.27}$$

The direction should be checked carefully. If L_{ij} is larger than L_{ij}^0, then the spring is stretched and will exert a force on object i toward object j, or in the same direction as the unit vector. The force exerted on object j is clearly the negative of Eq. 7.27.

The harmonic approximation is now made by making the assumption that the Δx_i's and Δy_i's are small compared to any of the L_{ij}^0's. Expanding Eq. 7.27 in powers of Δx_i and Δy_i gives

$$\begin{aligned}\vec{F}_i = K_{ij}([&\cos^2\theta_0(\Delta x_j - \Delta x_i) + \cos\theta_0 \sin\theta_0(\Delta y_j - \Delta y_i)]\hat{e}_x \\ &+ [\sin^2\theta_0(\Delta y_j - \Delta y_i) + \sin\theta_0 \cos\theta_0(\Delta x_j - \Delta x_i)]\hat{e}_y),\end{aligned} \tag{7.28}$$

where

$$\cos\theta_0 = \frac{|x_j^0 - x_i^0|}{L_{ij}^0} \tag{7.29}$$

and

$$\sin\theta_0 = \frac{|y_j^0 - y_i^0|}{L_{ij}^0}. \tag{7.30}$$

From Eq. 7.28 we see that the force that acts on object i due to all of the springs will always be a linear function of the Δx_i's and Δy_i's. If we use a $2N$-component vector to represent these displacements:

$$\underline{\Delta x} = \begin{bmatrix} \Delta x_1 \\ \Delta x_2 \\ \vdots \\ \Delta y_1 \\ \Delta y_2 \\ \vdots \end{bmatrix}, \tag{7.31}$$

then Eq. 7.28 can be written in terms of a dynamical matrix

$$\underline{F} = -\underline{\underline{\mathcal{D}}} \cdot \underline{\Delta x}, \tag{7.32}$$

and the analysis proceeds in an analogous manner to the one-dimensional case. Each normal mode will in general involve motions of the objects in both directions.

It is important to recognize that the harmonic approximation enters in two ways here. First, the assumption is made that the spring forces are linearly related to the stretch or compression of the spring, and second, the displacements are assumed to be small so that only terms that are linear in the displacement are kept in the expression for the force. This second assumption is not necessary in one dimension where ideal springs generate a harmonic system for any magnitude displacements. This small-displacement assumption has some important consequences. For example, the motion of an object in a direction perpendicular to the spring connecting it to another object does not generate any restoring force.

7.3 Using the Program

7.3.1 Structure of the Simulation

The COUPOSC Program allows the user to specify a configuration of objects and springs in one or two dimensions. The configuration may consist of up to 10 objects connected by as many as 45 springs. The mouse or the keyboard can be used to add and delete objects and springs and to drag objects to new locations. Three mass types and four spring types are available for constructing the configuration. An infinite mass type is also available that can be used to generate fixed points in the configuration. Note that the equilibrium length of each spring is determined by the equilibrium separation of the two objects it connects. It is not possible in this simulation to set up an equilibrium configuration with springs in a stretched state.* The numerical values of the three mass types and four spring types can be modified by the user.

Once the configuration has been specified, the program generates the dynamical matrix for the system, and then solves the eigenvalue problem (Eq. 7.13) for the normal mode frequencies and eigenvectors. The solution may include zero values for ω^2 and/or negative values for ω^2 (and thus imaginary values for ω). Zero values correspond to translational and/or rotational motion and will occur in one dimension if there is no infinite mass to tie the system down. In two dimensions, two infinite masses are required to hold the system in place and to prevent rotations. Negative values for ω^2 result from an insufficient number of constraints (i.e., springs), so that the equilibrium configuration is not a stable equilibrium. The simulation will allow investigation of systems with zero or imaginary eigenfrequencies, although these modes are not included in computing the motion of the system. The dynamical matrix, and the eigenvectors and eigenvalues, can be viewed from within the simulation. They are also written to the .CFG file when the configuration is saved.

After calculating the normal modes, the simulation displays a histogram giving the occupation of each normal mode. This histogram can be used to study the individual modes by occupying a single mode at a time and observing the vectors representing the relative displacements of each object. See Figure 7.1. Combinations of several modes may also be constructed. Phases of individual modes can

*This means that the problem of stretched strings cannot be investigated with this simulation. This problem is, however, considered in detail in the Waves and Optics CUPS simulation.

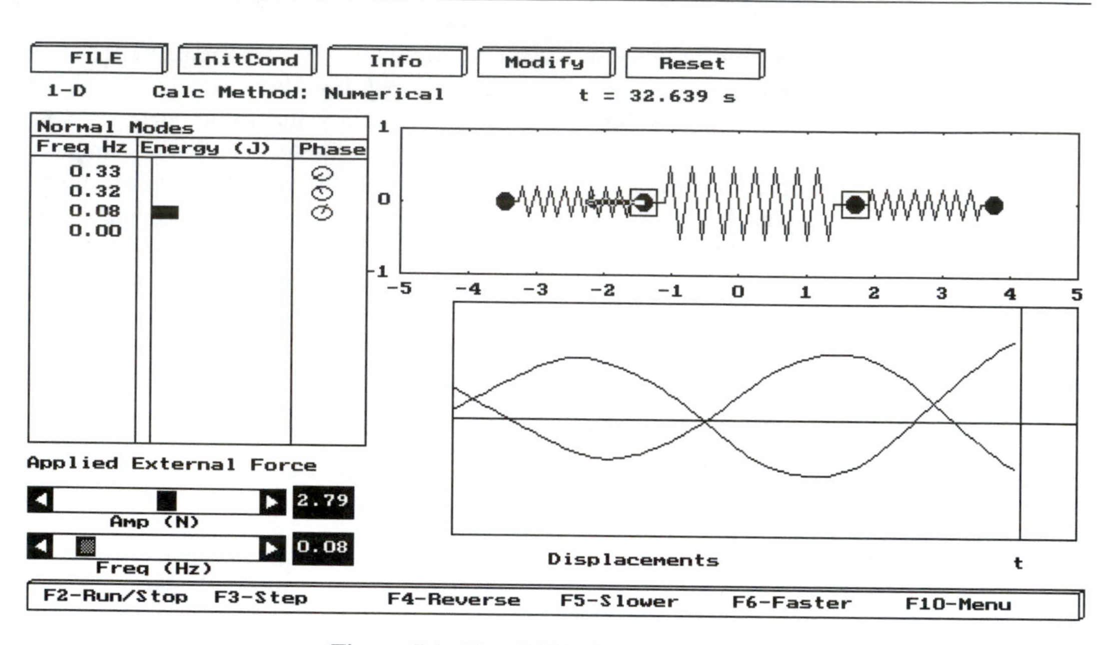

Figure 7.1: The COUPOSC user screen.

be toggled between 0° and 180°. Specific initial displacements may also be specified in the animation window. The normal mode occupation is dynamically updated whenever initial displacements are changed.

Note that while working with the normal mode histogram or when specifying initial displacements in the animation window, the initial velocities are set to zero. This is designed to avoid confusion between initial displacement vectors and initial velocity vectors. Initial velocities may be included by selecting the **Fine Tune Initial Displacements** menu option or by specifically setting phase values in the **Fine Tune Normal Modes** menu option.

When the initial conditions have been set, the subsequent motion can be generated using one of two options. The **Normal Modes** solution uses Eq. 7.15 to generate the time evolution of the object positions. The **Numerical** solution uses the fourth-order Runge-Kutta method to generate the positions, using the dynamical matrix to generate the forces applied to the objects. The **Normal Modes** solution method will be faster and is not subject to significant numerical approximation error. The **Numerical** solution method, however, allows the inclusion of additional external forces. The simulation includes a damping force that is linear in the object velocities with the same coefficient for all objects and a periodic driving term that can be applied to one of the objects. Exercises give examples of simple extensions to the simulation to allow for unequal damping, simultaneous application of driving forces to more than one object, and non-sinusoidal driving forces. When the numerical solution is run, the occupation of various modes can be dynamically displayed, so that phenomena such as the relative decay of various modes due to damping, and the effect of resonance of the driving force, can be observed. Thus, even when the forces are not strictly harmonic, the harmonic solution can provide valuable insight into the dynamics.

7.3.2 The User Interface

When **COUPOSC** is run, the default system is drawn on the right-hand side of the screen and a window is opened on the left displaying a histogram of the current normal mode occupations for the system. The hot keys at the bottom of the screen allow you to run the simulation to observe the time development of the system. The animation may be paused by hitting any key or clicking the mouse. The animation may be speeded up, slowed down, or run in reverse by hitting the various hot keys.

The various menu items in the simulation are described below. Note that detailed context-sensitive help is available by typing F1 at every step while running COUPOSC. The information provided in this section will be most helpful when you want to modify the program.

Program Menus

- **Important Preliminaries**
  You exit the program by selecting **EXIT Program** under the **File** option in the menu. Help can be obtained by typing F1. You may save a system you have constructed by selecting **Save System** under the **File** option in the menu.

- **File**
  This selection allows you to load and save files and exit the program. A system consists of the specific equilibrium arrangement of objects and springs and also the current initial conditions of the system. Thus, a particular set of interesting initial conditions can be saved and loaded later for presentations or copied to disk and turned in as part of a homework assignment. The dynamical matrix, eigenvalues, and eigenvectors are also saved with the system. These may be extracted from the system file for additional numerical work.

- **InitCond**
  This selection allows you to investigate the motion associated with individual modes and the effect of combining various modes. If you select **Fine Tune Normal Mode Occupations**, you can specify initial mode occupations and phases directly. A better way to investigate the individual modes, however, is to select **Modify Normal Mode Occupations**, which then allows you to change mode occupations directly on the histogram while vectors indicating initial displacements are drawn on the picture of the system. Note that when you select this option, all initial velocities are set to zero. This gives the clearest representation of the normal modes, since all the energy of a mode starts out as displacements. The **Fine Tune Initial Displacements** option allows you to specify the initial displacements and velocities. The normal mode occupations will be adjusted accordingly. The **Modify Initial Displacements** option allows you to change graphically the initial displacements while the normal mode occupations are adjusted dynamically. In this option, the initial velocities are set to zero so that normal mode occupations represent displacements only.

- **Info**
  This selection allows you to look at the numerical values of the dynamical matrix, eigenvectors, and eigenvalues. The **Show Mass/Spring Types** option

displays the current numerical values of the masses and spring constants that constitute the system. These values can be changed within the **Modify System** option. The **Show Cur Pos/Vel** option shows the current values of the displacements and velocities, and the **Show Mode Occupations** option shows the energy in each normal mode.

- **Modify**

  - **Modify Current System**
    This selection allows you to modify the arrangement of objects and springs that constitute the system. You can add, delete, or move objects, add or delete springs, and change mass and spring types. Remember that a system is built of various combinations of four spring types and three mass types, plus an infinite mass. You can change the numerical values of the three masses and the four spring constants by selecting **Modify Mass/Spring Values** from the menu. Note that the mouse can be used to select objects or springs or to reposition objects, but a mouse is not required to make changes.

  - **Modify Simulation Parameters**
    This selection allows you to modify various aspects of the simulation and display. The maximum mode energy represents the upper limit of the scale of the mode occupation histogram. Increase this energy if you want to be able to generate larger initial displacements.

    The animation interval is the default time step used between screen updates. Since the speed of the animation is limited by the time required for screen updates, increasing this parameter will increase the animation speed, but may also cause the motion to appear discontinuous.

    You may choose a numerical solution of the equations of motion, rather than the eigenvalue solution. This option uses the fourth-order Runge-Kutta method numerically to integrate the equations of motion. Use this numerical solution if you want to include non-harmonic forces such as friction, or a driving force. These forces are specified using the **Modify External Force** option. Note that you have an option of continuously updating the mode occupation. This allows you to watch the relative decay of various modes due to damping or the enhancement of particular modes when a resonant periodic force is applied. Several of the exercises investigate phenomena by the use of this option.

    In one-dimensional simulations, you have the option of displaying a dynamic graph of the displacement versus time of two of the objects in the system. You may change the objects that are graphed by checking the appropriate box in this input screen.

  - **Modify External Force**
    Use this option to modify the external (non-harmonic) forces applied to the system. Note that these forces are ignored unless the program is using a numerical solution (selected from the **Modify Simulation Parameters** option). A friction term that is linear in the velocity can be applied to all the masses and a periodic driving force can be applied to one mass. The magnitude and

frequency of this driving force can be modified while the motion is in progress, using the sliders that appear on the screen whenever the numerical solution method is selected. Note that the mass to which this force is applied can be changed during the simulation by hitting the space bar.

- **Reset**
  This menu option resets the system to the equilibrium state and resets the time to zero. To view the initial displacements, you may hit F3 to move the system to its initial position, or you may click the mouse in the animation window to see the vectors representing initial displacements.

7.4 Exercises

7.1 Two identical objects of mass m are constrained to move in one dimension and are connected by a spring with length L and spring constant k. Each object is also connected to a rigid support by a spring with spring constant K. See Figure 7.2.

(a) Set up the dynamical matrix for this problem and solve for the normal mode eigenvalues ω_a and ω_b and their corresponding eigenvectors in terms of m, k, and K. Describe the motion corresponding to each mode, and write down the general expression for the motion of each object, $\Delta x_1(t)$ and $\Delta x_2(t)$. Note that these expressions should have four free parameters to be determined by the initial conditions.

(b) Use the COUPOSC simulation to set up a configuration representing this system. Use $m = 0.5$ kg, $K = 20$ N/m, and $k = 10$ N/m. Show that your expressions for ω in part a generate the same values determined by the simulation. Note that the simulation reports actual frequencies, not *angular* frequencies. Check your descriptions of the normal mode motions.

(c) Determine $\Delta x_1(t)$ and $\Delta x_2(t)$ in terms of m, K, k, and L if object 1 is initially displaced a distance $L/8$ to the *left*. Note that you must do the algebra here, but you should check your results by considering the specific numerical values and comparing them to the simulation results. What is the ratio of λ_a to λ_b? What is the ratio of the energy in mode a to the energy in mode b? What is the effect on the initial conditions of changing the phase of mode a by 180° (note that you can do

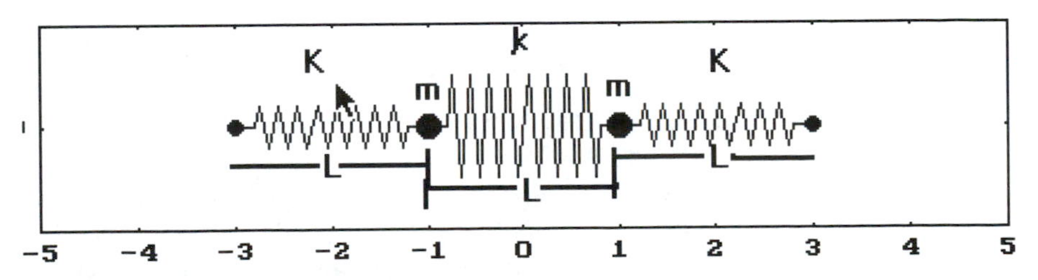

Figure 7.2: Exercise 1.

this in the simulation by clicking on the phase symbol in the **Mode Occupation** window)? What is the effect of changing the phase of mode b by 180°? Run the simulation with these initial conditions and describe the resulting motion of the two particles. Is the motion periodic?

(d) Suppose object 1 has zero initial displacement, but is given an initial velocity of v_0 to the right. Determine $\Delta x_1(t)$ and $\Delta x_2(t)$ in this case. What is the ratio of λ_a to λ_b? What is the ratio of the energy in mode a to the energy in mode b? Run the simulation with these initial conditions (try $v_0 = 5$ m/s) and describe the resulting motion of the two particles. How is the motion different from that in part c?

7.2 Consider the same system described in Problem 1. If you have not already done so, work out the normal modes of this system, and determine $\Delta x_1(t)$ and $\Delta x_2(t)$ in terms of m, K, k, and L if object 1 is initially displaced a distance x_0 to the left, as done in part c of Exercise 1. Now consider the case where k is much smaller than K, so that the two masses are only weakly coupled.

(a) Expand your expression for the larger of the two frequencies (call this ω_a) up to linear order in k/K. If you have access to a symbolic algebra program (e.g., Mathematica or Derive), keep terms up to third order. Use $m = 0.5$ kg, $K = 20$ N/m, and $k = 2$ N/m in the COUPOSC simulation and in your approximate expressions, and show that you get the same result within the limits of your expansion.

(b) Consider the motion of this system if object 1 is initially displaced a distance x_0 to the left while object 2 remains at its equilibrium position. Using the trigonometric identity for $\cos(A) + \cos(B)$ and, keeping only the linear term in the expansion for ω_a, show that the motion of object 1 can be written as

$$\Delta x_1(t) \approx x_0 \cos(\omega_b(k/2K)t) \cos\left(\frac{1}{2}(\omega_a + \omega_b)t\right), \qquad (7.33)$$

and show that this implies that the *amplitude* of oscillation of object 1 oscillates with a period equal to $(2K/k)T_b$, where T_b is the period at which object 1 would oscillate if the coupling were zero. Test this result by running the simulation and determining the time between points at which the motion of object 1 decreases nearly to zero.

(c) Use the simulation to investigate the effect of using objects with slightly different masses. Try masses that differ by 20 percent.

(d) Load the default one-dimensional (1-D) configuration in the COUPOSC simulation and investigate the transfer of energy between the two pairs of objects. What is the period of this energy transfer? Try to determine how this period depends on the spring constants. Do not do the algebra, but investigate how the period varies as the spring constants are varied.

7.3 **Molecular Vibrations**: Consider a linear, triatomic molecule such as CO_2 that has identical atoms with mass m at the ends, an atom with mass M at the center, and identical "spring constants," K, connecting the atoms.

(a) Set up the dynamical matrix for this problem and solve for the normal mode frequencies in terms of M, m, and K. Show that the ratio of the two non-zero frequencies is independent of the spring constant.

(b) The reported vibration frequencies of CO_2 are 4.010^{13} Hz and 7.0510^{13} Hz. Use your results from part a to determine the spring constant K that will generate these frequencies. Note: For this part, use MKS units for the masses and spring constant.

(c) In setting up a simulation for this molecule, it is very inconvenient to use the numerical values from part b because they are such small numbers. Chemists often use units where masses are measured in atomic mass units (AMU) and lengths are measured in angstroms (Å). Frequencies are often reported in cm^{-1}, which corresponds to the reciprocal of the wavelength of light with the given frequency. Using these units, determine the numerical value of the spring constant that will give the correct frequencies (in inverse centimeters). Use these numbers in the simulation to check that the frequencies you obtain are identical to the reported values in part b. What are the units of the spring constant? Chemists generally measure energies in kJoules/mole (kJ/mol), so that the spring constant would be measured in kJ/mol/Å.^{2} What is the value for K in these units?

7.4 In part a of the previous problem you showed that the ratio of the two normal mode frequencies was independent of the spring constant. This will be true for any system composed of a single type of spring.

(a) Set up a 1-D configuration of four identical objects of mass $m = 1$ kg, separated by equilibrium distance $L = 2$ m, and connected by three identical springs with spring constant $K = 15$ N/m. Record the normal mode frequencies and determine the ratios of each pair of eigenfrequencies.

(b) Change the spring constant to 30 N/m and record the new mode frequencies and ratios.

(c) Use dimensional analysis to show that the frequencies obtained in part b must be $\sqrt{2}$ times larger than the frequencies from part a. Hint: The algebraic equations for the frequencies will undoubtedly be fairly complicated, but you know that they can only contain the parameters K and m, and there is only one way to combine these parameters to obtain a frequency. What effect would changing all of the masses by a factor of 2 have on the frequencies?

(d) Repeat parts a and b using a different mass at the ends of the chain. Try $M = 2$ kg for the end masses and $m = 1.0$ kg for the center masses. Can you predict what effect this change will have on the frequencies? Is the scaling of the frequencies with the spring constant still the same?

(e) Load the default 2-D configuration in the COUPOSC simulation and investigate the scaling of the normal mode frequencies with the spring constants. Modify the configuration so that all three objects have different masses, and investigate the scaling again.

7.5 **Resonance**: Set up a 1-D configuration consisting of five equally spaced objects with 2 m between each pair. The objects on the ends should have infinite mass and the remaining three should have masses of 1.0 kg. Each adjacent pair of objects is connected by identical springs with spring constant equal to 25 N/m.

(a) Determine the normal mode frequencies of this configuration. Record these frequencies and describe the motion of each mode.

(b) Select the **Numerical Solution Method** under the **Modify Simulation Parameters** menu option. Select the the **Modify External Force** menu option and set the frequency of the external force to match the lowest eigenfrequency and include a frictional coefficient of 0.5 N/(m/s). Apply a 5.0 N force to the leftmost object. Run the simulation for a while to allow the system to pass through the "transient" period and determine the energy stored in the lowest frequency mode. Note that you can determine this energy by pausing the motion and using the **Show Mode Occupations** option under the **Info** menu option.

(c) Increase the frequency of the external force by a small amount and repeat the determination of the maximum amplitude described in part b. Investigate a range of frequencies to determine the dependence of occupation energy on external frequency. Draw a graph of energy versus frequency. The *width* of the resonance is defined as the shift in driving frequency that generates a mode occupation energy equal to one half the energy exactly at resonance. Determine this resonance width.

(d) Determine the width of the other two resonances. The *quality*, Q, of a resonace is defined as the ratio of the resonance width to the resonance frequency, $Q = \Delta\omega/\omega$. Which of these resonances has the largest Q?

(e) Change the force so that it is applied to the middle mass. What happens when the frequency of the driving force is equal to the second normal mode frequency in this case?

7.6 **Sound Waves in a Solid**: Set up a 1-D configuration consisting of nine identical objects, each of mass 10 kg, connected by identical springs of spring constant, 20 N/m. Arrange the objects symmetrically in the window with a distance of 1.0 m between each pair.

(a) Investigate the motion corresponding to each normal mode by occupying that mode and observing the initial displacements and subsequent motion. Make a graph of frequency versus mode number starting with the lowest frequency as mode 1.

(b) Set up the initial conditions so that only the leftmost object has a displacement to the right (use **Modify Initial Displacements** under **InitCond**). Record the occupation of each mode corresponding to this initial condition (use **Info/Show Mode Occupations**). Can you account for the symmetries of these occupations? Run the simulation to determine how long it takes for this disturbance to reach the right-hand side of the system (stop the clock when the right-hand mass first begins to move). Calculate the "wave speed" by dividing the distance traveled (8.0 m) by this time. Does this speed depend on the amplitude of the initial displacement? Change the value of the spring constant and repeat the experiment. Determine how the speed depends on the spring constant. Change the numerical value of the mass and repeat the experiment. Determine how the speed depends on the masses. Increase the distance between objects by removing every other object. How does this affect the wave speed? Do these results agree with your intuitive understanding of the speed of sound waves?

(c) Set up the initial conditions so that the two leftmost moveable objects have equal displacements to the left. Give the object at the end a displacement about twice that of the second object. How is the mode occupation different for this set of initial conditions? Run the simulation to determine the speed of this wave (remember to set the mass back to 10 kg and the spring constant back to 20 N/m). Is it the same as the speed found in part b (you might argue that the distance the wave must travel in this case is only 7.0 m)?

(d) Change every other object in this chain, starting with the leftmost mass, to a second mass type. Set the mass value of these objects to be 12 kg and change the original mass type to 8.0 kg. How are the frequencies different from those found in part a? How are the normal mode displacements different? Plot these frequencies on your graph of frequency versus mode number. Change the mass values again to 18 kg and 2.0 kg for the two types and include these frequencies on your graph. You should find that the frequencies have split into two separate bands. What happens when the second mass type becomes infinite?

7.7 **Percussion Instruments**: Set up a 1-D configuration consisting of nine objects, each of mass 0.03 kg and equally spaced. Connect adjacent pairs with identical springs of spring constant, 20 N/m. Change the object at each end to an infinite mass. This system represents a simple model of a xylophone bar, although vibrations in this system are longitudinal, whereas the xylophone bar vibrates primarily in transverse modes.

(a) The musical *quality* of an instrument usually depends on the extent to which the various vibration modes that are excited form a *harmonic series*, that is, whether the normal modes can be written as integers times a fundamental frequency. How close are the normal mode frequencies to a harmonic series? What is the fundamental frequency of this system?

(b) In a real xylophone, the wooden bars are shaped to cause the frequencies to resemble more closely a harmonic series. Usually this requires reducing the mass of the wood near the center of the bar. Try adjusting the masses in your model so as to generate a harmonic series more closely. Record the mass values you use for each object and plot the frequencies as a function of mode number to show whether your series is *harmonic*. Warning—musical instrument builders have spent centuries perfecting their craft. Don't expect perfection.

(c) Change all the masses back to 0.03 kg. What is the period of the highest frequency normal mode? Use the numerical solution method and set the maximum of the external force to 15 N and the frequency of this force to zero. Set the damping coefficient to 0.04 N/(m/s). Apply the force to the leftmost moveable object (use the spacebar to move the applied force to a different object). Step the simulation through a time period equal to one half the period of the highest frequency normal mode, then set the force to zero (use the slider) to simulate striking the instrument with a hammer. Record the occupation of the modes at this point. How is the energy distributed in these modes? Continue to run the simulation. What happens to the energy in the normal modes? Do all modes lose energy at the same rate? Will the *quality* of the sound change as the signal weakens?

(d) Repeat part c, but this time allow the force to be applied for the full period of the high frequency mode. How is the occupation of the modes different for this case? Use this result to explain why a soft hammer (which remains in contact with the block for a longer time) generates a more mellow tone.

(e) Repeat part c with the force applied to the middle mass (use a one-half period contact time again). How is the occupation of the modes different for this case? Demonstrate why certain modes should be absent in this case. Why is it important musically where the block is struck?

(f) **Program Modification**: Modify the program so that the external force is applied to two objects instead of one. This can be done most easily by adding a few lines of code in the Procedure **GetAccel** that calculates the force acting on each object. The number of the mass you wish to use can be determined by using the **Info/Show Current Pos/Vel** menu option. Note that this will not cause the second force vector to be drawn on this mass. That can be accomplished if you wish by directly modifying the **DrawExtForce** procedure to draw both vectors. Repeat part c with the force applied to the two leftmost objects (use a one-half period contact time again). How is the mode occupation different for this case? Why is the size of the hammer important in determining the quality of sound obtained from a xylophone?

References

1. Press, W. H., Flannery, B. P., Teukolsky, S. A., Vetterling, W. T. *Numerical Recipes in Pascal*. Cambridge: Cambridge University Press, 1989.

2. Arfken, G. *Mathematical Methods for Physicists*. San Diego: Academic Press, Inc., 1987.

Appendix A

Walk-Throughs for All Programs

These "walk-throughs" are intended to give you a quick overview of each program. Please see the Introduction for one-paragraph descriptions for all programs.

A.1 GENMOT

You can run a few of the sample GENMOT programs (PENDULUM, 1DGRAV, RACETRACK, etc.) if you *really* want to, but you **should** run the Borland Pascal editor, BP, and create your own. I describe setting up a simple pendulum below.

- **Load a Program:** Load 1DGRAV.PAS into the BP editor and change the filename to YPF (your personal filename).
- **Define Variables in DefForceParms**
  - Change the text in **ForceTitle** and **ForceDesc**.
  - Change the dynamical variable names in **GenDynVar** to theta, omega, and alpha. Change units to radians.
  - Change the force parameters to m, g, and L with appropriate units and delete the fourth call to **GenForceParm**.
  - Rename the **ConfigFileName** to YPF.PRM.
- **Put the Differential Equation Into CalcAccel**
  - Change **x, vx, ax** in the procedure parameter list to theta, omega, alpha.
  - Change **m, g, c1, c2** in the procedure parameter list to **m, g, L, P4** (note that **P4–P10** are not used).

– Rename the variable **Force** to **Torque**.

– Insert the lines

Torque = −m*g*L*sin(theta);
alpha = Torque/(m*L*L);

- **Define the Dynamical Functions in CalcDynFun**

  – Make the same changes to the parameter list for this procedure (you can copy them from the **CalcAccel list**).

  – Change the kinetic energy to be m*SQR(L*omega)/2

  – Change the potential energy to be −m*g*L*cos(theta).

  – Fix up the total energy.

- **Compile the Program and Go...**

  – Make certain to enter a non-zero value for the mass and the length.

  – Set a non-zero value for the initial displacement (use **Initial Conditions** under the **Parameters** menu) and check to see that the motion looks reasonable and that the total energy is conserved (click on the numerical window at the top of the screen to change what is displayed).

- **Add an Animation of Your Pendulum**

  – Define two new dynamical functions xPos, yPos by including two more calls to **GenForceParm** in the Procedure **DefForceParm**. Add these parameters to the parameter list in **CalcAccel** and **CalcDynFun**.

  – Add the following definitions in **CalcDynFun**:

```
ELSE IF FunName = 'xPos' THEN
    CalcDynFun = L*sin(theta)
ELSE IF FunName = 'yPos' THEN
    CalcDynFun = -L*cos(theta);
```

  – Modify the parameter list of the Procedure **AnimateWindow** and add the lines

```
IF (yAxisVar = 'yPos') AND (xAxisVar = 'xPos') THEN
    PlotLine(0, 0, L*sin(theta), −L*cos(theta));
```

- **Run the Program:** Run the program and change one of the windows to display yPos versus xPos (click in the window you want to change). To change the sizes, positions, and number of windows, select **Modify Window Layout** under the **Windows** menu option.

A.2 ANHARM

The initial screen comes up with the mass in the center of a spring between two walls. Two graphs are shown, v vs. x (phase diagram) and x vs. t, and the motion starts automatically. Click the mouse button to stop it.

- **Click in the Phase Diagram.** The motion will start and run for two cycles with initial conditions determined by the position of the mouse cursor.

- **Move the Upper Wall.** Drag it with the mouse or use the F4 key until the potential vs. x diagram shows a double well. Click with the cursor in the potential diagram below the bottom of the hump separating the two wells. Repeat with the cursor on the other side of the center. Repeat with the cursor well above the hump and finally with the cursor just *slightly* above the hump so that the oscillator just barely makes it from one well to the other.

- **Chaotic Motion.** Select the menu item **Forces**, then **Interesting Parameters**, then **Two Walls**, then **Chaotic**. After watching the motion for a while, stop it and select **Display**, then **Display Two Graphs**, then F4 and F5. Now run it again for a while, then select **Choices** and **Poincaré Movie**. Notice the folding and layering in the movie. After that select **Frequency Analysis** from the **Choices** menu.

- **Long Period Motion.** Repeat the previous item with the **Long Period Interesting Parameter** choice. Compare the following x, v initial conditions: (0.5, 0), (0.75, 0), (0, 0.2), and (0 0.4). Notice that the Poincaré movie does not show the folding and layering that the chaotic one did.

- **Short Period Motion.** Compare the following initial conditions: (1, 0) and (−1, 0).

- **Period.** Select **Plot Period** from the **Choices** menu.

- **Pendulum.** Select **Forces**, **Pendulum**. Click near 3 rad on the x-axis, then near 8 rad/s on the v-axis. Try turning the animation on again if you turned it off earlier (to be found on the **Choices** menu.)

- **Try Out the Tutorial** (on the **Help** menu).

A.3 ORBITER

The initial screen comes up with the entire solar system on the left, and a magnified view of just the inner planets on the right.

- **Press F2–Go to Start the Motion.** And F3–**Stop** to stop it.

- **Sun, Jupiter, and Comet.** Select this from the **Systems** menu. Notice that when the comet approaches the Sun the computation slows down to preserve precision. Let it run long enough for the comet to approach Jupiter and have its orbit changed. Now select **Replay** and notice that the comet speeds up when it should, near the sun.

- **Lagrangian Points.** Let this system run for a while, and observe the different behaviors of the three small bodies. Then choose **Restart** from the **Choices** menu, and choose **Replicate a Body** from the **Choices** menu. Replicate the body at 180° two times in the x direction (you will be asked for each of these choices). Run the system. Restart it again (the replicated system when asked) and from the **Plots & Zoom** menu choose two windows: the **Rotating** window and the **Second Rotating** window. Then **Zoom In** one of the windows on the replicated body. Launch the motion and enjoy.

- **Binary Stars & Comet.** Run the system for a while. Then experiment with moving the comet (**Move Body** on the **Choices** menu). Try adding another one.

- **Shuttle Docking.** See if you can dock the shuttle with the space station.

A.4 ROTATE

The initial screen comes up with a fixed axis system. Click on the **run** hot key or type F2 to see the simulation.

- **Adjust Object Orientation**. Change the initial conditions by selecting the **Parameters** menu option. Change the orientation of the angular velocity to phi = 180, theta = 60. Why does the torque disappear. Exercise 1 in this chapter will give you some practice with Euler angles.

- **Show Body-Centered Frame**. Select **Display** from the menu and include the body-centered frame. Rerun the simulation.

- **Use Stereographic Projection**. Select **Display** from the menu and include the stereographic projection. Try to make this work (I find about 50 percent success rate with this.)

- **Zero Torque**. Load the constant torque system by selecting **System** from the menu and note that the angular velocity always lies in the plane defined by the angular momentum and the z-axis.

- **Spinning Top**. Load the spinning top and let it run. Try decreasing the magnitude of the initial angular velocity.

A.5 COLISION

The initial screen comes up with a window to display trajectories in the laboratory system on the left and on to display trajectories in the center-of-mass system on the right. The former also displays a graph of the Yukawa potential, as noted above the latter.

- **Click Anywhere in the Left-Hand Window.** A collision will be displayed whose impact parameter corresponds to the location of the mouse pointer. Repeat as often as you like; use the F5–**Clear** key to clear the screen.

- **Change the Display.** Select F3–**Go New**, accept the **Single Collision** default choice, and select **Show Paths Only**. Click on **OK** or press **Enter** and the collision will run. Select several more impact parameters with the mouse.

- **Display Momentum Vectors as Well.** From **Plots & Zoom**, select **Trajectories & Vectors**. Run some collisions as before.

- **Comparing Cross Sections.** Select **Pure Coulomb** from the **Forces** menu, and accept the default **Repulsive**. Select F3 and choose **Plot Cross Section**. Change the number of points to 20, check **Heavily Concentrated**, and check **Save**. Press **Enter** and type a description, such as "Pure Coulomb, 20 points," then press **Enter**. Repeat with **Coulomb With Shielding** and **Coulomb Truncated**. If you don't get asked for the description, you forgot to check **Save**. Then choose **Display Saved Cross Sections** from the **Plots & Zoom** menu. Look at both the values and the logarithms.

- **Varying Parameters.** Select **Pure Coulomb** and F3. Check **Vary Energy**, then **Vary Force Strength**.

- **Try Out the Tutorial** (on the **Help** menu).

A.6 COUPOSC

The initial screen comes up with a one-dimensional example of a system that passes energy between two weakly coupled oscillators. Click on the **run** hot key or type F2 to see the simulation.

- **Adjust Normal Mode Occupations.** Click in the mode occupation diagram. Change the occupation of individual modes with the mouse or by using the keyboard. Note that vectors on the actual system give initial displacements of the objects. Look at each mode individually.

- **Adjust Initial Positions.** Click in the window containing the oscillating system and drag the objects to new initial positions. Note the changing mode occupations.

- **Apply External Driving Force.** Select **Program Parameters** and change to a numerical solution. Click in the mode occupation window and set all occupations to zero. Start the animation and watch the mode occupation change. Experiment with frequencies on and off resonance. Change the force to a different object by hitting the space bar.

- **Modify System.** Select **Modify System** from the menu and add some additional objects with springs.

- **Run a 2-D System.** Click on **File** and load the default 2-D system or the user-defined system, CRYSTAL.

Index

Limited Use License Agreement

This is the John Wiley and Sons, Inc. (Wiley) limited use License Agreement, which governs your use of any Wiley proprietary software products (Licensed Program) and User Manual (s) delivered with it.

Your use of the Licensed Program indicates your acceptance of the terms and conditions of this Agreement. If you do not accept or agree with them, you must return the Licensed Program unused within 30 days of receipt or, if purchased, within 30 days, as evidenced by a copy of your receipt, in which case, the purchase price will be fully refunded.

License: Wiley hereby grants you, and you accept, a non-exclusive and non-transferable license, to use the Licensed Program on the following terms and conditions only:

a. You have been granted an I [illegible] single personal computer for your own personal use only.
b. A backup copy or copies m [illegible] ns of this agreement.
c. You may not make or distr [illegible] on, or otherwise, the source code of the Licens [illegible] or in part, except as expressly permitted by th [illegible]
d. A backup copy or copies [illegible] rmitted by this Agreement.

If you transfer possession of a [illegible] se is automatically terminated. Such termination shall be in a [illegible] Wiley.

Term: This License Agreemer [illegible] the Licensed Program with all copies made (with or without [illegible]

This Agreement will also term [illegible] il to comply with any term or condition of this Agreement. [illegible] pies made (with or without authorization) of either.

Wiley's Rights: You acknowl [illegible] secrets) in the Licensed Program (including without limitation, [illegible] d all means and forms of operation of the Licensed Pro [illegible] nent, you do not become the owner of the Licensed Progra [illegible] Agreement. You agree to protect the Licensed Program from u [illegible] he Licensed Program contains valuable trade secrets and con [illegible] ent of the Licensed Program, whether or not in machine rea [illegible]

THIS LIMITED WARRANT [illegible], INCLUDING WITHOUT LIMITATION, ANY WARR [illegible] RPOSE.

EXCEPT AS SPECIFIED AI [illegible] S IS" BASIS AND WITHOUT WARRANTY AS TO THE I [illegible] D PROGRAM. THE ENTIRE RISK IS TO THE RESULTS [illegible], REPAIR, OR CORRECTION OF THE LICENSED PROGR [illegible]

IN NO EVENT WILL WILEY BE LIABLE TO YOU FOR ANY DAMAGES, INCLUDING LOST PROFITS, LOST SAVINGS, OR OTHER INCIDENTAL OR CONSEQUENTIAL DAMAGES ARISING OUT OF THE USE OR INABILITY TO USE THE LICENSED PROGRAM EVEN IF WILEY OR AN AUTHORIZED WILEY DEALER HAS BEEN ADVISED OF THE POSSIBILITY OF SUCH DAMAGES.

THIS LIMITED WARRANTY GIVES YOU SPECIFIC LEGAL RIGHTS. YOU MAY HAVE OTHERS BY OPERATION OF LAW WHICH VARIES FROM STATE TO STATE. IF ANY OF THE PROVISIONS OF THIS AGREEMENT ARE INVALID UNDER ANY APPLICABLE STATUE OR RULE OF LAW, THEY ARE TO THAT EXTENT DEEMED OMITTED.

This Agreement represents the entire agreement between us and supersedes any proposals or prior agreements, oral or written, and any other communication between us relating to the subject matter of this Agreement.

This Agreement will be governed and construed as if wholly entered into and performed within the State of New York.

You acknowledge that you have read this Agreement, and agree to bound by its terms and conditions.